PROBABILISTIC THINKING, THERMODYNAMICS, AND THE INTERACTION OF THE HISTORY AND PHILOSOPHY OF SCIENCE

VOLUME II

VOLUME 146

PROBABILISTIC THINKING, THERMODYNAMICS AND THE INTERACTION OF THE HISTORY AND PHILOSOPHY OF SCIENCE

Proceedings of the 1978 Pisa Conference on the History and Philosophy of Science

VOLUME II

Edited by

JAAKKO HINTIKKA, DAVID GRUENDER

Dept. of Philosophy, Florida State University, Tallahassee

and

EVANDRO AGAZZI

Dept. of Philosophy, University of Genoa, Italy

SPRINGER-SCIENCE+BUSINESS MEDIA, B.V.

Library of Congress Cataloging in Publication Data

CIP

Pisa Conference on the History and Philosophy of Science,
1978.
Probabilistic thinking, thermodynamics, and the interaction of the history and philosophy of science.

(Its Proceedings of the 1978 Pisa Conference on the History and Philosophy of Science; v. 2) (Synthese library; v. 146)
Includes index.
1. Science—History—Congresses. 2. Science—Philosophy—Congresses. 3. Thermodynamics—Philosophy—Congresses. 4. Probabilities—Congresses.
I. Hintikka, Kaarlo Jaakko Juhani, 1929–
II. Gruender, C. David, 1927–
III. Agazzi, Evandro, 1934– IV. Title.
Q124.6.P57 1978, vol. 2 509s [509] 80–22437
ISBN 978-90-481-8361-6 ISBN 978-94-017-2766-2 (eBook)
DOI 10.1007/978-94-017-2766-2

TABLE OF CONTENTS

SECTION VI: WHAT CAN THE HISTORY AND PHILOSOPHY OF SCIENCE AND PHILOSOPHY DO FOR EACH OTHER? A ROUND-TABLE DISCUSSION

TABLE OF CONTENTS OF COMPANION VOLUME

SECTION III: THE PHILOSOPHICAL PRESUPPOSITIONS AND SHIFTING INTERPRETATIONS OF GALILEO

PREFACE

The two volumes to which this is a preface consist of the Proceedings of the Second International Conference on History and Philosophy of Science. The Conference was organized by the Joint Commission of the International Union of History and Philosophy of Science (IUHPS) under the auspices of the IUHPS, the Italian Society for Logic and Philosophy of Science, and the Domus Galilaeana of Pisa, headed by Professor Vincenzo Cappelletti. Domus Galilaeana also served as the host institution, with some help from the University of Pisa. The Conference took place in Pisa, Italy, on September 4–8, 1978.

The editors of these two volumes of the Proceedings of the Pisa Conference acknowledge with gratitude the help by the different sponsoring organizations, and in the first place that by both Divisions of the IUHPS, which made the Conference possible. A special recognition is due to Professor Evandro Agazzi, President of the Italian Society for Logic and Philosophy of Science, who was co-opted as an additional member of the Organizing Committee. This committee was otherwise identical with the Joint Commission, whose members were initially John Murdoch, John North, Arpad Szabó, Robert Butts, Jaakko Hintikka, and Vadim Sadovsky. Later, Erwin Hiebert and Lubos Novy were appointed as additional members.

The announced program of the Conference is included in each of the two volumes. A number of changes as compared with this planned schedule are worth mentioning. A reception in honor of the Conference was given by the city of Pisa on September 6, and is hereby gratefully acknowledged. Several of the expected participants were unable to come to Pisa. They include Erwin Hiebert, A. T. Grigorian, Wesley C. Salmon, M. Clavelin, Karel Berka, I. A. Akchurin, B. M. Kedrov, V. S. Kirsanov, and S. R. Mikulinsky, (Arpad Szabó participated only in the first two and a half days of the meeting.) Of the absentees, Hiebert, Salmon, Clavelin, Kirsanov (jointly with Markova), and Berka have nevertheless contributed to the proceedings of the Conference. As will be seen from these volumes, interesting contributions were also made by originally unscheduled participants. L. Jonathan Cohen's contribution to the Pisa Conference has meanwhile appeared in a modified form in the *Journal of the History of Ideas* **41** (1980), 219–231. The modifications have rendered redundant the interesting comments which Robert Butts contributed to the Conference. This is not the case with Ian Hacking's remarks in our Vol. II below, which now should be referred to Cohen's paper, published as indicated above, even though Professor Hacking, through no fault of his own, wrote his contribution on the basis of what Dr. Cohen said in Pisa.

As is seen from the program below, the Conference was organized in six sections. Papers from Sections I–III are published as Volume I, and those from Sections IV–VI as Volume II of these Proceedings.

The papers published in these two volumes speak for themselves. They are not primarily intended to present the last word on any one subject. Rather, their purpose is not just to contribute to a discussion of a worthwhile subject which is of interest to both historians of science and philosophers of science, but also to exemplify the actual co-operation of representatives of these two sister disciplines. The co-operation of those two fields of study has in fact long been one of the aims of Synthese Library. Hence it was not hard to find a natural publication forum for these proceedings, especially as the output of its predecessor conference in 1964 has been partly published in the same series. It is not for the managing editor of Synthese Library to judge the papers appearing here in the light of this objective of the Conference. It is in any case amply obvious that the Conference produced extremely lively exchanges and several unexpected combinations of ideas.

Another aim of the organizers was to have a variety of approaches and a variety of geographical areas represented at the Conference. Whether this aim was realized or not as fully as it could have been, we are in any case pleased to be able to include in this volume several contributions by participants from Eastern Europe which will allow the readers of Synthese Library an interesting glimpse of the work done there in the field of the Conference. Again, the aims of the organizers and the policies of Synthese Library converge.

There could not have been a more felicitous location for a joint meeting of philosophers and historians of science than Domus Galilaeana. The organizers are most grateful to Professor Cappelletti for this privilege. It was symbolically appropriate that the first volume of the *Carteggio dei Discepoli di Galileo* was officially presented to the public as a part of the Conference. In a wider sense, we are all disciples of Galileo, even though these Proceedings appear in Synthese Library and not in the *Carteggio.*

We are thankful to D. Reidel Publishing Company for including these two proceedings volumes in one of their series. Much of the detailed editorial work was done by Ms. Jayne Moneysmith, to whom our warm thanks are hence due. In the compilation of the indexes, she was assisted by Mr. Jim Garrison and Ms. Susan Leffler, whom we would also like to thank.

On behalf of the editors and organizers,

JAAKKO HINTIKKA

PROGRAM OF THE SECOND INTERNATIONAL CONFERENCE ON THE HISTORY AND PHILOSOPHY OF SCIENCE

Pisa, Domus Galilaeana: 4–8 September 1978

OPENING ADDRESS: Vicenzo Cappelletti

TOPIC I:	The Structure of Theory Change
FOCUS:	The Sneed–Stegmüller "Rational Reconstruction" of Kuhn's Views
CHAIRMAN:	Lorenz Krüger
PAPERS BY:	Ilkka Niiniluoto (Finland) – Zev Bechler (Israel)
COMMENTATORS:	Joseph Sneed (Holland) – Vadim Sadovsky (USSR) – Paolo Rossi Monti (Italy)

TOPIC II:	The Early History of the Axiomatic Method
FOCUS:	Szabo's Theory of the Origin of the Axiomatic Method
CHAIRMAN:	John Murdoch
PAPERS BY:	Wilbur Knorr (USA) – Patrick Suppes (USA) – M.V. Popovich (USSR)
COMMENTATORS:	Árpád Szabó (Hungary) – Karel Berka (Czechoslovakia)

TOPIC III:	Philosophical Presuppositions and Shifting Interpretations of Galileo
FOCUS:	The Shift in the Interpretation of Galileo from Koyré to Drake
CHAIRMAN:	Vincenzo Cappelletti
PAPERS BY:	David Gruender (USA) – W. Wisan (USA) – V.S. Kirsanov (USSR) – L.A. Markova (USSR)
COMMENTATORS:	M. Clavelin (France) – Carlo Maccagni (Italy) – A.C. Crombie (GB) – N. Jardine (GB) – B. Kouznetsov (USSR)

TOPIC IV:	Probability Theory and Probabilistic Thinking in the Seventeenth to Nineteenth Centuries
FOCUS:	Hacking's View on the Development of Probability Concepts
CHAIRMAN:	Robert Butts
PAPERS BY:	Ivo Schneider (BRD) – Anne Fagot (France)
COMMENTATORS:	Ian Hacking (Canada and USA) – L. Jonathan Cohen (GB) – A.T. Grigorian (USSR) – V.N. Kostjuk (USSR) – Lubos Novy (Czechoslovakia)
TOPIC V:	Thermodynamics and Physical Reality since 1850
FOCUS:	The Mach–Boltzmann Contrast
CHAIRMAN:	Patrick Suppes
PAPERS BY:	Erwin Hiebert (USA) – Lorenz Krüger (BRD) – Ulises Moulines (Mexico)
COMMENTATORS:	John North (GB and Holland) – Nancy Cartwright (USA) – I.A. Akchurin (USSR) – Kurt M. Pedersen (Denmark) – E. Bellone (Italy)
TOPIC VI:	Round Table Discussion: WHAT CAN THE HISTORY AND PHILOSOPHY OF SCIENCE DO FOR EACH OTHER?
CHAIRMAN:	Evandro Agazzi
	John Murdoch (USA) – Robert Butts (Canada) – Erwin Hiebert (USA) – B.M. Kedrov (USSR) – S.R. Mikulinsky (USSR) – Evandro Agazzi (Italy) – A.T. Grigorian (USSR)

The first volume of the *Carteggio dei Discepoli di Galileo* will be presented during the course of the meeting (Thursday 7 September, 6 p.m.)

DIRECTORS	SECRETARIAT
Jaakko Hintikka	Guido Cimino
Evandro Agazzi	Marcello Tricarico
Vincenzo Cappelletti	

SECTION IV

PROBABILITY THEORY AND PROBABILISTIC THINKING IN THE CLASSICAL MODERN PERIOD

IVO SCHNEIDER

WHY DO WE FIND THE ORIGIN OF A CALCULUS OF PROBABILITIES IN THE SEVENTEENTH CENTURY?*

Whenever one asks why something like a calculus of probabilities arose in the 17th century, one already assumes several things: for instance, that before the 17th century it did not exist, and that only then and not later did such a calculus emerge. If one examines the quite impressive secondary literature on the history of probability, one finds that it is by no means a foregone conclusion that there was no calculus of probabilities before the 17th century. Even if one disregards numerous references to qualitative and quantitative inquiries in antiquity and among the Arabs and the Jews, which, rather freely interpreted, seem to suggest the application of a kind of probability-concept or the use of statistical methods, it is nevertheless certain that by the end of the 15th century an attempt was being made in some arithmetic works to solve problems of games of chance by computation. But since similar problems form the major part of the early writings on probability in the 17th century, one may be induced to ask why then a calculus of probabilities did not emerge in the late 15th century.

One could say many things: For example, that these early game-calculations in fact represent one branch of a development which ultimately resulted in a calculus of probabilities. Then why shouldn't one place the origin of the calculus of probabilities before the 17th century after all? Quite simply because a suitable *concept* of probability was missing from the earlier computations. Once the calculus of probabilities had been developed, it became obvious that the older studies of games of chance formed a part of the new discipline.

We need not consider the argument that practically all the solutions of problems of games of chance proposed in the 15th and 16th centuries could have been viewed as inexact, and thus at best as approximate, by Pascal and Fermat in the middle of the 17th century, – that is, before the emergence of a calculus of probabilities.

The assertion that no concept of probability was applied to games of chance up to the middle of the 17th century can mean either that

J. Hintikka, D. Gruender, and E. Agazzi (eds.), Pisa Conference Proceedings, Vol. II, 3–24.

there existed no concept of probability (or none suitable), or that though such a concept existed it was not applied to games of chance. I consider the latter to be correct, and in this I differ from Hacking, who argues that an appropriate concept of probability was first devised in the 17th century.[1]

I should like to mention that Hacking and I agree on a number of points. For instance, on the significance of the legal tradition and of the practical ("low") sciences: Hacking makes such factors responsible for the emergence of a new concept of probability, suited to a game calculus, while I perceive them as bringing about the transfer and quantification of a pre-existent probability-concept.

To defend my thesis I shall first show that such a concept did in fact exist. I shall then explain why it remained impossible to connect the concepts of chance and probability until 17th century. As the final step I shall portray the background which made possible a quantification of the probable.

The antecedent concept I have in mind is not to be understood as an extension of a medieval scholastic view of probability, such as one meets in Thomas Aquinas,[2] but is rather a direct adoption of one already fully elaborated in antiquity. "Direct" is to be taken literally, i.e., such pioneers in the quantification of the probable as Leibniz and the authors of the Logic of Port Royal drew not from a continuing tradition, but rather direct from ancient Greek sources, indeed frequently with verbatim citations. Thus, to answer why, no quantification of probability was attained earlier when such a concept had been so long at hand, I will propose that this quantification required a displacement or, better, an extension of this concept's realm of application and that this occurred only in the 17th century.

It is true that Aquinas' understanding of probability, which Hacking contrasts to the later concept, stands in a tradition which extends back into antiquity, but this takes into account only a part of the ancient understanding. According to Byrne the spectrum of meanings which Thomas assigned to "*probabilis*" coincides perfectly with Aristotle's use of "*endoxos*". By "*endoxos*" he denoted the range of a belief dependent on the state of information of the subject; for Aristotle this precursor of a later concept of subjective probability was significant above all in dialectical debate and particularly in rhetoric, so important for political and juristic decision-making. (It is interesting that these areas of application of the probability concept became of explicit concern in this same context once again in the 17th & 18th centuries.) In contrast to Hacking, who views Aristotle and his followers as irrelevant because of their great chronological remoteness, one can show that the understanding of probability in the 17th century was linked directly with that of antiquity. For instance, Leibniz refers not only to Aristotle's *Topics*, where the meaning of "*endoxos*" is explained,[3] but also directly to the phrase "*hōs epi to poly*"[4], which Aristotle uses in anticipation of an understanding of probability independent of the subject and which for a given starting situation describes the subsequent situation that as a general rule is to be expected.

Aristotle took this concept of the general rule, that is, of that which occurs in most or the majority of cases, from the field of medicine. For example, in the first book of the Hippocratic tract *De Morbis* one finds repeatedly the formulation that a sickness of such-and-such a kind will "in most cases" end in death. Aristotle links the notion of the general rule with "*endoxos*" because the subjective expectation of an event which occurs as a rule is greater than that of the exception to the rule. It is thus decisive for the realm of applicability of the Aristotelian understanding of probability that, of the three realms of (1) that which always and therefore necessarily happens, (2) that which happens in most cases or as a rule and (3) that which happens by pure chance, only the first two are accessible to scientific research. Since one may further show that in Aristotle's view games of chance belong to the third realm, it follows that, according to his dogma, probability, taken as a scientific notion cannot be applied to them. The impact of this in the 17th century can be seen in Johannes Kepler, who never applies the concept of "*probabilitas*" in his deliberations on dice and the realm of chance.[5]

Up to now we have considered an understanding of probability only as it appears in the thought of Aristotle. But we have yet to examine the elaboration of this Aristotelian understanding by the Sceptics of the Middle Academy, in particular by Carneades. From the writings of Sextus Empiricus one learns that Carneades distinguished three levels of the probable as the only possible landmarks for one's decisions in practical life and that each of these levels embraces a continuous interval of intensities.[6]

Even if the achievements of the Sceptics had no influence, they would be of interest in their own right. To be sure, virtually nothing of Sceptical thought passed into the scholastic tradition. But from the middle of the 16th century the writings of Sextus Empiricus became available again, in printed form. Furthermore, and above all in view of the then-current theological controversies, the need for coming to terms with Scepticism had long since become clear.[7]

In particular one can discover a familiarity with Sceptical ideas taken from a reading of Sextus in Leibniz and in the Logic of Port Royal. Hence, a concept of probability suited to the quantification of the probable achieved in the second half of the 17th century could have been derived directly from the Sceptical tradition.

At the same time one sees that, as long as the Aristotelian dogma of the impossibility of any scientific treatment of chance held sway, games of chance could not be included within the realm of "*probabilis*"; and moreover that any attempt to mathematize problems about games of chance within the framework of the established scientific domain was inconceivable. The mathematization of games of chance had to be achieved in an environment freed from the tabus of scholastic science. Such an environment, denoted by Hacking as "low sciences", was defined by the maestri d'abaco (reckoning-masters), who above all satisfied the need for training in the economic and commercial spheres of the late middle ages, a training not available within the universities.

The significance of commerce for the rise of games of chance seems so evident to Hacking that he fails to pursue it. Of course, on the basis of his thesis, that a concept suited to the quantification of probability had first to be created, Hacking must likewise turn to areas outside the established domain of science; but his motives for doing this are entirely different from my efforts to substantiate attempts to solve problems of games of chance numerically on the basis of the economic realities at the conclusion of the economic revolution of the Middle Ages.

From the above it is clear that such an attempt could be undertaken only in a community which found itself outside the dominion of the Aristotelian dogma; that is, it either knew nothing of the alleged impossibility of treating chance by scientific means, or else it could afford to ignore it.

A Church teaching was also overthrown in the course of the economic expansion of the late Middle Ages and early modern period: the idea that money is sterile. This notion staunchly defended by theologians, was weakened by among other things the establishment of games of chance in commercial circles and by the attempt to view determinate situations in games of chance as analogous to risk-taking ventures in trade. One example of this is the structural equivalence between the problem of dividing stakes when a game must be interrupted (problem of points) and the division of gains and losses in the so-called "*commenda*". (Contracts like the "*commenda*" completely disregarded the Church's insistence on the sterility of money.) Games of chance, which of course have always existed, were spread far and wide from the 14th century on by the travelling merchants, and came to be seen as temporally and spatially condensed representations of commercial activity. Above all one saw an essential parallel in the chance redistribution, of an investment (stake) among the participants. It is not surprising, then, that the determination of profit shares that was a matter of course in high-risk trading ventures, particularly in the sea trade, was transferred to gaming situations, in which risk enters in the form of the unknown outcome of a chance event.

Besides the need to supervise affairs of trade through the fullest possible quantification of all relevant factors it was necessary to consider the dimension of *time*, especially in speculative business. Contrasting with the medieval world-view, the new sense of man as largely responsible for the events of this world and therefore as a *planning* individual is of particular significance for the early forms of commercial forecasting, out of which there developed a general interest in future events. Connected with this new interest is the goal of controlling the future.[8]

In the aftermath the efforts to solve problems of games of chance in the Italian arithmetics up to the middle of the 16th century were interrupted, largely because the Counter Reformation re-emphasized traditional moral values and not only prohibited games of chance but also regarded them as snares of the devil. This is confirmed by the fact that virtually nothing was published in this field until the correspondence between Pascal and Fermat and the publication of Huygens' tract on games of chance. Almost a century later, and that even in the second half of the 17th century some works on games of chance were published anonymously.

At the French court, the rage for games of chance, imported from Italy, took hold only in the 17th century. In the middle of the century there grew the feeling, at least in Paris court circles, that the tabus of the Church were no longer binding as they had been a century before. The development of the new mathematics by Viète and Descartes seemed at the same time to justify an enthusiastic optimism that all conceivable problems could be solved mathematically. An oral tradition, going back to the attempts of the Italian maestri d'abaco of the late 15th and 16th centuries, may have contributed to the extension of this optimism to the realm of chance. This led finally to the well-known successes of Pascal, Fermat and Huygens in solving special problems of games of chance by the methods of the new algebra. Christiaan Huygens represents the best example of this development. Having heard in Paris of Pascal's and Fermat's successful efforts in solving such problems but without being able to obtain details about the basic fundamental approaches and methods employed, he nevertheless set his small but historically most significant treatise concerning the determination of values of expectation on a basis which corresponds entirely to commercial thinking. This is revealed in his central concept "*kans*" (Latin: "*expectatio*") as well as in the principle he states at the beginning of the "just game", (lusus justus) from which he derives his numerical values for expectation. He also expressed openly his astonishment at the power of the "analysis", by which he means the letter-algebra established by Viète and Descartes, in a letter to Frans van Schooten published together with his treatise, and viewed his own work as an example of a new field of application, that of chance events, uncertain in their outcome. He writes,

> The more difficult it seemed to determine through reason what is unsure and subject to chance, the more remarkable should that science appear through which such a thing is subjected.[9]

The analogy between trade and games of chance, so obvious at the time, made it easy for Huygens to use the value of expectation as his central concept. But having achieved the numerical determination by means of algebra, he was left facing a problem of justification. After what we have said above it is not surprising that words like "*probabilis*" or "*verisimilis*" nowhere appear in Huygens' treatise. These concepts, which fall within the tradition of the Aristotelian "*endoxos*", he uses in an entirely different context, for example in his discussion of the possibility of life on other planets.

Relatively soon after the first publication of Huygens' treatise we encounter an extension of the validity of "*probabilis*" to chance events like games of chance. But before I set out the reasons for this, I wish briefly to survey the contents of the sixteenth and last chapter of Part IV of the *Ars cogitandi* by Arnauld and others, in which the ratios of the chances for specific events are equated with the ratios of the associated degrees of probability. This chapter stands at the end of a discussion about judging the certainty of occurrence of events. This discussion fits directly into the Sceptical tradition. After renouncing the dogmatic claim to a true and immutable Being, the Sceptics of the Middle Academy strove constructively to offer rules for conduct in daily life; this conduct was to be guided by decisions based on the subjective evaluation of probabilities of the occurrence of events. While in the 15th chapter the *Ars cogitandi* discusses the evaluation of our belief in past events, in the sixteenth chapter it takes up the belief in future contingent events. Already in the thirteenth chapter a rule of conduct is proposed that appears to have been drawn directly from the Sceptics: to judge whether or not one should believe in the occurrence of an event, one must not consider the event in isolation, as if it were a geometric theorem, but rather one should take into account all the circumstances, both internal and external. The more detailed exposition of the circumstances to be considered agrees with that of Carneades in the case of the idea he designates as "*indubitable*". To this effect the sixteenth chapter of the Logic of Port Royal,[10] which treats of future events, recommends that the following principle be applied:

> Just as we ought to believe in an event if certain circumstances are known to have occurred which are commonly connected with the occurrence of such an event, so also if the present circumstances are those ordinarily followed by such an event, we should believe that the event will probably happen. According to this rule, doctors judge of the favorable or unfavorable issue of a disease, generals about the future events of a war, and ordinary men daily judge of the greater part of contingent affairs.[11]

The attendant observation that, in the decisions of daily life, only the gain hoped for or the loss feared are generally taken into account, without consideration of the *probability* that gain or loss will occur, motivates a discussion of degrees of probability. As a model for the quantitative discrimination of such degrees of probability the example of a game involving ten players with equal chances of winning and equal contributions to the stakes is used. The winner takes all, i.e., nine others besides his own stake, while the other players lose theirs.

This situation is explained as follows: "Thus, a player has nine silvers to hope for, one piece to lose – nine degrees of probability of losing a coin, and only one degree of winning nine".[12] The only thing new here, in comparison with Huygens' computation of expectation values, is the translation of ratios of chances in random decisions into a ratio of degrees of probability.

To readers of the widely-disseminated, Jansenist *ars cogitandi* this translational equivalence – prompted by the subject of game-calculations, future events with uncertain outcomes – may have suggested that the entire realm to which the concept *probabilis* was applicable could be made accessible to mathematics. The *Ars cogitandi* nevertheless gives us no indication of how one is to evaluate the circumstances relevant in any particular case to an event of interest.

As factors promoting the use of the ratio of chances model in determining the ratio of degrees of probability one might cite the following:

1. A new attitude toward the future as an outgrowth of economic speculation. In the 17th century it had developed into the quest for rational planning; with this was joined the question of the "certainty" with which one could expect the occurrence of future events.

2. A concept of chance much changed from the "blind chance" of the Epicureans and from the Aristotelian concept. A new understanding of chance arose through the notion of being placed in a world determined by divine Providence; it allowed the application of "*probabilis*" to chance events.

3. The revival of Sceptical ideas, through which the idea of degrees of probability, as had been distinguished in the Middle Academy, were once again brought into currency.

4. The struggle against the probabilism of the Jesuits, with the aim (among other things) of hindering the misuse of authority with the help of quantitative approaches in the realm of probabilities hitherto only qualitatively differentiated. The extension of the applicability of "*probabilis*" to future chance events suggested the adoption, for the purposes of quantification, of the games of chance model which had developed independent of any understanding of probability.

Beyond this by no means insignificant suggestion, the *Ars cogitandi* contains little of consequence for the foundation of a calculus of probability. The same is also true for another example, to which Hacking assigns great weight: the investigations of Gottfried Wilhelm

Leibniz, especially those concerning the evaluation of the then so-called conditional law (*jus conditionale*). One can in fact find passages in the work of Leibniz beyond those cited by Hacking that establish Leibniz' interest in making the evaluation of degrees of probability in adjudication exact. The significance of jurisprudence as an area for the application of the calculus of probability developed by Jakob Bernoulli is indicated not only in the correspondence of Bernoulli and Leibniz, but also in the dissertation of Niklaus Bernoulli, Jakob's nephew, who sought to apply the findings of his uncle to a series of concrete problems in law. Hacking holds the view that the 1660's were the decisive period in the development of the numerical evaluation of probabilities, to support which he cites Leibniz and the *Ars cogitandi* as principal witnesses. In contrast to this position, I maintain that the decisive program of finding a measure for probability conceived of as a degree or fraction of certainty and erecting upon this measure the first calculus of probabilities was carried out by Jakob Bernoulli, who took from his predecessors essentially little more than an inspiration.

The decisive works of Bernoulli, dating from the 1680's, applied concrete computation to problems other than the calculation of games of chance and of mortality. The role played by Leibniz must now be considered once again, in order to assess the achievement of Bernoulli in becoming the true creator of the calculus of probabilities.

I will thus seek to answer the following questions:

1. Were Leibniz' ideas on probability novel?
2. Why was he interested in the evaluation of probabilities?
3. What examples of concrete numerical evaluation of degrees of probability are to be found in Leibniz' works?

One should first note, as Hacking correctly observes, that Leibniz intended to devise a new logic for determining different degrees of the probable even before his stay in Paris, that is, at a very early stage of his scientific development. For instance, in his letter of December 18/28, 1670, Friedrich Nitzsche urges Leibniz to work out his announced "*Doctrina de gradibus probabilitatis*"[13]; this establishes that Leibniz had in mind the program of setting out a more comprehensive "*Doctrina*" quite early.

In the autumn of 1679, thus after his Parisian stay, Leibniz wrote to Duke Johann Friedrich of a "Nouvelle logique, pour connoistre les degrés de la probabilité", and states that such an "art" had not yet been developed, although it would be of great importance, in particular for the courts.[14]

Concerning the question of Leibniz' sources for his understanding of probability and of the distinction of degrees of probability, one may observe that for the then usual understanding *probabilitas* he turns directly to Aristotle, whom he frequently cites, especially in connection with the *Topics.* Of greater interest here, however, are his references to the Middle Academy and the Sceptics, for these suggest that Leibniz owed to his study of the Sceptics' ideas (as available from Sextus Empiricus) his distinction of degrees of probability. A number of reasons may be proposed to explain why Scepticism gained currency among natural scientists in the second half of the 17th century, Robert Boyle being perhaps the clearest example. I will mention here only the problem of the quest for an authority at a time when the absolute claims to truth and authority of the Bible and the Catholic Church were being challenged by the similar claims of natural philosophy and the Protestant Church.

That Leibniz' notion of probability is similar to that of the Greeks can be shown from passages in his *Nouveaux essais sur l'entendement humain.* These studies, a commentary on John Locke's *Essay concerning Human Understanding* (1690), were written at a time when Leibniz had come to realize that he himself could no longer carry out his own early program of a logic for the evaluation of different degrees of probability. In these *Nouveaux essais,* not published during his lifetime, Leibniz took up the epistemological problem (Book IV), within which he considered the notion of degrees of assent (ch. 16). One cannot fail to recognize the debt to Sceptical thought in this chapter.[15]

It is also clear from this work that Leibniz, in his distinction among the different degrees of probability, was able to specify concretely only qualitative differences, not levels which might be conceived quantitatively. For instance, in Section 5 he writes:

Pour revenir au fondement et aux degrés de l'assentiment, il est à propos de remarquer que les propositions sont de deux sortes: les unes sont de fait, qui dependant de l'observation peuvent estre fondées sur un temoignage humain; les autres sont de speculation, qui, regardant les choses que nos sens ne scauroient nous decouvrir, ne sont pas capables d'un semblable temoignage.[16]

In Section 9 he then elaborates:

Mais lorsque les temoignages se trouvent contraires au cours ordinaire de la nature, ou entre, eux, les degrés de probabilité se peuvent diversifier à l'infini, d'où viennent ces degrés que nous appellons *croyance, conjecture, doute, incertitude, defiance*; et c'est là où il faut de l'exactitude pour former un jugement droit et proportionner nostre assentiment aux degrés de probabilité.

Th. Les Jurisconsultes en traitant des preuves, presomtions, conjectures et indices, ont dit quantité de bonnes choses sur ce sujet, et sont allés à quelque detail considerables.[17]

Here, as in the examples from civil law which follow, Leibniz is thinking not of a numerical gradation of the probable, but of a qualitative ordering, corresponding to the view of the Sceptics, and indeed they too had entertained the notion of a continuum of possibilities of modality. His comment in the same paragraph on the situation in medicine confirms this:

Les Medecins encor ont quantité de degrés et de differences de leur *signes* et *indications*, qu'on peut voir chez eux. Les Mathématiciens de nostre temps ont commencé à estimer les hazards à l'occasion de jeux.[18]

Leibniz was already long familiar with the connection between ratios of chances and probability when he wrote this; at the same time he was corresponding with Bernoulli on probability. From these passages it is clear that he still hoped that mathematicians might realize his youthful program. In this sense he expresses himself once again at the end of the same paragraph:

J'ay dit plus d'une fois qu'il faudroit une *nouvelle espece de Logique*, qui traiteroit des degrés de probabilité, puisqu' Aristote dans ses Topiques n'a rien moins fait que cela, et s'est contenté de mettre en quelque ordre certaines regles populaires, distribuées selon les lieux communs, qui peuvent servir dans quelque occasion, où il s'agit d'amplifier le discours et de luy donner apparence, sans se mettre en peine de nous donner une balance necessaire pour peser les apparences et pour former là dessus un jugement solide. Il seroit bon que celuy qui voudroit traiter cette matiere, poursuivit l'examen des *jeux de hazard*; et generalement je souhaiterois qu'un habile Mathematicien voulût faire un ample ouvrage bien circonstancié et bien raisonné sur toute sorte de jeux.[19]

We have now seen that Leibniz' understanding of probability was linked directly with that of the ancients and that his motives for attempting to evaluate degrees of probability grew chiefly out of his study of law. It remains to inquire whether he did not after all go on to make quantitative trials in this field. To this end, we may consider two questions: (1) What was Leibniz' own judgment of such efforts? (2) Did Leibniz give examples of evaluations of probabilities?

Of interest in connection with the first question is a passage in a letter Leibniz wrote to Gabriel Wagner (1698), in which he says that in determining the "*gradus probabilitatis*", one ought to pay attention to

> wie man die anzeigungen, so keinen vollkommenen Beweiß machen und gegen einander laufen (indicantia et contra-indicantia, wie die Medici reden) abwegen und schäzen solle, umb den außschlag zu geben. Denn man insgemein gar wohl sagt, rationes non esse numerandas sed ponderandas, man müße die anzeigungen nicht zehlen, sondern wägen, aber niemand hat noch dazu die Wage gezeiget, wiewohl Keine dem werck näher gekommen und mehr hülffe an hand gegeben als die Juristen, daher ich auch der materi nicht wenig nachgedacht, und dermahleins den mangel in etwas zu ersezen hoffe.[20]

From this passage it is clear that Leibniz saw the best start towards the realization of his program among the jurists. His hope to compile his own reflections remained unfulfilled at his death in 1716. From this same passage it is also clear that he regarded the problem of evaluating degrees of probability as no mere enumeration, say of the circumstances favoring the occurrence of an event, but rather as a weighting that allows for the diversity of these circumstances. This does not mean that he excludes the usual enumeration of "equipossible" cases, as done in deriving the ratios of chances in computing games of chance; on the contrary, an example in another passage, having to do with the relative evaluation of events, namely, of obtaining a 9 or a 7 by the cast of two dice, shows how such a weighting can succeed, at least in computing games of chance.

It would now be interesting to see in what way, if at all, Leibniz succeeded in making such evaluations in the area of law. Hacking has already examined his treatment of "*jus conditionale*", which, in contrast to "*jus nullum*" (assigned the value 0) and "*jus purum*" (assigned the value 1), is given a fractional value between 0 and 1. Unfortunately, examples in which some such fractional value is actually assigned in a concrete instance of law are absent from the extant juristic writings of Leibniz. In the second part of the paper *De conditionibus* he presents a fictitious example of the relative ordering of competing claims: one person, under condition *A*, will receive 100 talers, under condition *B*, 200 talers. With the further assumption that the occurrence of both conditions is equally uncertain, he says that the "*jus conditionale*" dependent on *B* ought to receive a higher value than that under A.[21]

In like manner Leibniz stated in the *Specimina juris* (1669) the general principle that the greater the probability a condition exists, the higher the value of the "*jus conditionale*" associated with it. He adds that this probability increases with the number of possibilities for the existence of the conditions in question.[22]

Finally, one finds in a draft of the *Elementa juris naturalis*, dating from the second half of 1671, an example of a decision effected via computation. It is to be decided whether one should choose action A or action B after considering the probability that an effect resulting from the action will occur and the bearing of the associated effect. The interpretation of *effect* as *utility* motivated Leibniz' principle of decision-making, namely, to choose that action for which the *product* of the probability and the utility is a maximum. From the arbitrary values which he introduced into his example one notes that the probabilities used are not normalized. Thus, there is no question here of evaluating a concrete instance, but only of purely fictitious figures.

All this demonstrates that Leibniz failed to carry out his own program of evaluation concretely, both in his published works and in his posthumous and unpublished papers. He held essentially qualitative distinctions among degrees of probability, as was then common in the practice of law and medicine. But he maintained that his program could be implemented by a mathematician who studied systematically and in detail problems of games of chance and of games in general.

Leibniz lived to hear of such a mathematician, who independent of him and without knowledge of his program had worked on the application of an "*ars conjectandi*", as this mathematician called the numerical calculation of probabilities, to "civic, moral and economic matters". This man was Jakob Bernoulli, whose investigations came to Leibniz' attention only shortly before Bernoulli's death. An assessment by Leibniz of Bernoulli's achievement indicates clearly who, in Leibniz' own view, had founded the calculus of probabilities and thereby made possible, to use one of Leibniz' own phrases, a new logic that treats of degrees of probability (*logique, qui traiteroit des degrés de probabilité*). In his letter to Bourguet of March 22, 1714, Leibniz emphasized that Bernoulli had cultivated the art of conjecture on the basis of his own recommendations, which he in turn had based on the start made by Pascal, Fermat, Huygens, de Witt and Hudde.[23]

In this passage Leibniz also refers to a distinction Bernoulli made between the determination of probabilities *a priori* and *a posteriori.*

That Bernoulli conducted his investigations on the basis of Leibniz' recommendations is not merely an exaggeration; it is demonstrably false. Leibniz' influence on the *Ars conjectandi* was confined to a few passages in the fourth part, where Bernoulli takes up certain objections which Leibniz had expressed in correspondence. But the remark in the letter to Bourguet reveals Leibniz' concern to assure himself some credit for the creation of the *Ars conjectandi*, and thus how highly he esteemed this first work on the calculus of probability.

Bernoulli left the *Ars conjectandi* as an unfinished manuscript, whose content in its most important sections went back to preliminary studies he had done in the 1680's. These early studies are now available, with the publication of the relevant passages of his scientific diary, the *Meditationes.*[24] One can now reconstruct the origins of his various ideas on probability. One can identify with certainty among his sources, first, the tract by Huygens, which was reprinted in the first part of the *Ars conjectandi* with Bernoulli's annotations; the *Ars cogitandi*; and the combinatorial investigations by Pascal in his *Triangle arithmétique.* One can see how Bernoulli, beginning from the notion, contained in the Logic of Port Royal, of the identification of ratios of chances with the ratio of degrees of probability, developed the classical concept of probability, still current far into the 19th century, and how he established as its measure a generalization of Huygens' determination of expectation, namely, the ratio of favorable to possible cases. At the same time he became the first to set down the prerequisites for consciously formulating a program for the mathematization of all the fields of application subject to "*probabilis*". Bernoulli himself sought to execute this program, but his premature death prevented him. His nephew, Niklaus Bernoulli, who was especially close to him in the last years, took up the work of applying the *ars conjectandi* to the study of law, where in contrast with Leibniz' efforts concrete instances of law were treated numerically. To be sure, this development belongs to the early part of the 18th century. But the emergence of Jakob Bernoulli's ideas falls entirely within the 17th.

A key passage for the transformation in the conception of probabilities and of probability involves the treatment of a problem of law, on which Bernoulli worked, according to the *Meditationes*, in 1685–

86. It has to do with a marriage contract, which, assuming that the couple is blessed with children and that the wife dies before the husband, will govern the division of their common property between the father and the children. A distinction is made among the possibilities that both, one or neither of the fathers of the bridal couple, alive at the time of the conclusion of the contract, die and leave their estates to their children. Only the property which has become the common property of the couple is to be regulated, and so such distinctions need to be drawn. The portion of the groom will be larger if he has already entered into his inheritance, smaller if not, unless both fathers have died. The bride's father objects to this initial proposal; this induces Titius to make a second proposal, according to which he will receive the same portion of the common property regardless of what happens to the fathers.

On this basis Bernoulli poses the question: which suggestion would be more favorable for the children? To this end he has to make assumptions about the possible orders of death of the three people involved, the two fathers and the bride, Caja. He first assumes that all six possible orders have equal weight. But this assumption does not satisfy him, since the youth of the bride has not been taken into account. Thus, he assumes that for every two instances – e.g., diseases, symptoms or events – which might bring about the death of either father, there is only one which threatens Caja with death. There are thus five cases in all, each equally likely to take its victim first. Since Caja is affected by only one of these, while the two fathers are affected by four, her situation is evaluated as one-fifth of certainty of her being first to die, that is, "one probability, five of which make the entire certainty".[25]

Here Bernoulli uses the plural "probabilities", where these are equated with the no more precisely distinguished individual cases; this usage does not yet permit the conception of "probability" as "degree of certainty" which is observed in the next stage. Aided by Huygen's formula for determining expectation, Bernoulli then derives a certainty of $\frac{4}{15}$, written $\frac{4}{15}c$ (where "c" stands for "certitudo"), or 4 probabilities out of 15, that Caja will die second, and finally $\frac{8}{15}c$ that she will die third. (We will not here describe the further hypotheses which Bernoulli employs to weight the various orders to death).

It is interesting that at this time Bernoulli appears not to have had knowledge of either the *Observations* of Graunt (1662) or the *Waer-*

dye (*Estimate*) of de Witt (1671).[26] Yet at the end of his treatment of the marriage contract between Titius and Caja he proposes extensive investigations on human mortality. Here he speaks also of the degree of probability:

Generally in civic and moral affairs ⟨things⟩ are to be understood, in which we of course know that one thing is more probable, better or more advisable than another; but by what degree of probability or goodness they exceed others we determine only according to probability, not exactly. The surest way of estimating probabilities in these cases is not *a priori*, that is, by cause, but *a posteriori*, that is, from the frequently observed event in similar examples. In our example, if in the course of many years it had been observed that twice as many old men died as young girls, all of whom were of the same age and constitution as our young girls and old men, we would conclude then that there was one case which threatens the young girl with death and two cases which threaten the old man; or if among various pairs of young and old men it had been observed that the young man outlived the old man paired with him in twice as many cases as those in which the old man outlived the young man, we would conclude on the other hand that there is one case which deprives the young man of life before the old man and two cases which produce the opposite.[27]

Just before this passage Bernoulli had introduced the fundamental distinction between determination *a priori* and *a posteriori*:

The reason, that in card and dice games, which are governed solely by chance, the expectation can be precisely and scientifically determined, is that we can perceive accurately and clearly the number of cases in which gain or loss must follow infallibly and that these cases behave indifferently and can each occur with equal facility or when one is more probable than another we can at least determine scientifically by how much it is the more probable. But what mortal, I pray you, counts the number of cases, diseases or other circumstances to which now the old men, now the young men are made subject, and knows whether or not these will be overtaken by death, and determines how much more probable it is that one will be taken unawares than another, since all of these depend on causes that are completely hidden and beyond our knowledge.[28]

One should regard the somewhat pessimistic conclusion of this passage in the light of the passage which follows it and has been cited above, in which the execution of probabilities *a posteriori* is explained.

How does Bernoulli intend to carry out the determination of probabilities *a posteriori*? In particular, does he suppose that the reliability of the derived values will increase with the number of observations? Bernoulli sought to answer these questions by means of what he called his golden theorem (theorema aureum), the first version of which appeared about 1689. In the time between this version and his earlier treatment of the problem of the marriage contract, he came to avoid using the plural "probabilities" in the sense of the different equipossible cases in games of chance. In formulating the theorem he used "*probabilitas*" only in the singular, in the sense of the degree of certainty with which the occurrence of an event may be expected. What he establishes in this theorem is that as the number of observations of a repeatable event increases, so too does the probability that the relative frequency of occurrence of a possible outcome will lie in the vicinity of the probability of this outcome. Only much later did it become clear that he did and could prove this theorem only for the relative frequency of events of known *a priori* probability, but not for those of unknown probability. He understood this theorem as a justification for adopting relative frequencies determined through observation as estimates of probabilities which could not be given *a priori*. At the same time the fundamental theorem served as the essential foundation of Bernoulli's program to extend the realm of application of numerically determinable probabilities. Substantially because of this function of the theorem – which also became the starting-point for the calculus of probability established in the *Ars conjectandi* – its first formulation is given here in its simplest and still heuristic form:[29]

I can deviate from the true proportion less if I observe more often rather than less often.

E.g., when ⟨in a game⟩ there are as many chances for me as against me, equality in the ratio of games I have won to those my opponent has won is to be expected. Let a ratio be set which differs a little from that of equality, let us say 2 to 1; then I say the probability that the ratio of games ⟨won to those lost⟩ will deviate more from the ratio of equality is less when we play 6 games than when we play 3 and even less when we play 9 or 12 etc. Contrariwise, the probability is greater that ⟨the number of games won⟩ will be contained within three (sic) limits when we play more games than when we play fewer. Calculating:

When	3 games are played, to win	3,	2,	1,	0							games
	the number of cases is	1,	3,	3,	1							a total of 8.
For	6 games to win	6,	5,	4,	3,	2,	1,	0				games
	the number of cases is	1,	6,	15,	20,	15,	6,	1				a total of 64.
For	9 games to win	9,	8,	7,	6,	5,	4,	3,	2,	1,	0	games
	the number of cases is	1,	9,	36,	84,	126,	126,	84,	36,	9,	1	a total of 512.
For	12 games to win	12,	11,	10,	9,	8,	7,	6,	5,	4,	3,	
			2,	1,	0							games
	the number of cases is	1,	12,	66,	220,	495,	792,	924,	792,	495,	220,	a total of
			66,	12,	1							4096.

Therefore the probabilities that I shall win more than $\frac{2}{3}$ or fewer than $\frac{1}{3}$ of all the games are

for 3 games $\frac{1}{8}$
for 6 games $\frac{7}{64}$
for 9 games $\frac{46}{512}$
for 12 games $\frac{299}{4096}$.

But $\frac{1}{8} > \frac{7}{64} > \frac{46}{512} > \frac{299}{4096}$, which confirms etc.

Thus occurs what is usefully and joyfully to be observed, that by increasing the number of games one arrives at a probability that is smaller than any given probability; since the ratio of the individual ⟨probabilities⟩ to the following ones is larger than the ratio 8 to 7.

Therefore I should be able to observe so often that I ⟨shall be able to⟩ infer ⟨the ratio of chances⟩, with a probability bordering on certainty, as though I had this ⟨probability⟩ *a priori*. Query: How many observations must be carried out so that it will be 100,000 times more probable that the ratio of games I win to those I lose will approach that of equality more strongly than 101 to 99?

In effect, this heuristic discussion indicates that the probabilities, under the assumption that all chances are equivalent, that more than $\frac{2}{3}$ or less than $\frac{1}{3}$ of all the games will be won for a number $3n$, $n = 1, 2, 3, \ldots$ of games, are majorized by the sequence of general term $(7^{n-1})/8^n$, which tends to zero as n increases. Of particular interest is Bernoulli's use of normalized probabilities in such expressions as "the probability is $\frac{1}{8}$".

The most general form of the fundamental theorem was rigorously proved by Bernoulli only a short time later in the *Meditationes*. There it is presented in a form that holds for arbitrary initial probabilities and that is terminologically oriented toward calculating games of chance:

It is possible to carry out so many observations that, with any given high degree of probability, it will be more probable that ⟨the ratio⟩ of games won by both sides will lie within any given narrow limits rather than outside them.[30]

A last essential point for the research program of the "calculus of probabilities", the new discipline Bernoulli prepared for in the *Meditationes* and formulated in the *Ars conjectandi*, was a new concept of chance. To be sure, the latter work had assumed an understanding of the contingent that on the one hand permitted the application of the probable to chance events, and on the other hand merely asserted the compatibility of this concept of chance with divine Providence, without explaining this further. Bernoulli sought to close these gaps in the *Ars conjectandi*:

Contingent (in the sense "free"; insofar as it depends on the will of a rational creature, and in the sense "fortuitous" and "casual" insofar as it depends on a chance event or on fortune) is that which could not be, become or have been... Contingency does not always entirely exclude necessity, as far as secondary causes, as I shall make clear from examples.[31]

These examples make clear that Bernoulli never thought of events as occuring indeterminately. He was convinced that through a more precise knowledge of the parameters affecting the motion of a die, for instance, it would be possible to specify in advance the result of the throw. In similar fashion he viewed changes in weather as a determinate process, just as the occurrences of astronomical events are. Chance, in his view and later in the view of Laplace, was reduced to a subjective lack of information. Thus, depending on the state of their information, an event may be described by one person as chance, but by another as necessary. With this anticipation of Laplacian determinism Bernoulli appears to solve the problem of the connection between chance and divine Providence. The entire realm of events which are described in daily life as uncertain or contingent in their outcome is such, he claims, merely because of incomplete information; nevertheless, these too fall within the field of the concepts "*probabilis*". Bernoulli's program to mathematize as much of this realm as possible with the aid of the classical measure of probability occupied researchers throughout the 18th century and into the second half of the 19th.

NOTES

* Translation by W. Knorr, with assistance from D. Sepper.

[1] Hacking, I. *The Emergence of Probability: A Philosophical Study of Early Ideas of Probability*, Cambridge, 1975, see for instance p. 9 and in chapter 2 p. 17. As a philosopher Hacking tries to detect the modern concepts of probability in the 17th century, where he finds them "already full-fledged".

[2] Cf. Edmund F. Byrne, *Probability and Opinion: A study in the Medieval Presuppositions of Post-Medieval Theories of Probability*, The Hague, 1968, p. 188.

[3] Cf. *Nouveaux essais sur l'entendement humain*, IV, c. II, §14.

[4] See Leibniz' letter to James Bernoulli of December 3, 1703; cf. *Leibniz Mathematische Schriften*, vol. II, p. 84.

[5] Cf. Ivo Schneider, 'Wahrscheinlichkeit und Zufall bei Kepler', *Philosophia Naturalis* **16**, (1976), 40–63.

[6] Cf. Ivo Schneider, 'Contributions of the sceptic philosophers Arcesilas and Carneades to the development of an inductive logic compared with the Jaina-Logic', *Indian Journal for the History of Science* **12** (1977), 173–180. The relevant contributions of Aristotle are also reviewed in this article.

[7] Cf. Richard H. Popkin, *The History of Scepticism: From Erasmus to Descartes*, New York, 1964.
[8] For the connection between the economic development of the late Middle Ages and the computational treatment of games of chance, see Ivo Schneider and Karin Reich, 'Die wirtschaftliche Entwicklung des Mittelalters im Spiegel der arithmetischen Aufgabensammlungen und ihrer Nachfolger, der Rechenbücher des 15. und 16. Jahrhunderts', in *Aus dem Antiquariat*, June 1978, pp. 217–229; see also Ivo Schneider, 'Die Mathematisierung der Vorhersage künftiger Ereignisse in der Wahrscheinlichkeitstheorie vom 17. bis zum 19. Jahrhundert', in *Berichte zur Wissenschaftsgeschichte* 2 (1979), 101–112.
[9] "Quantò enim minus rationis terminis comprehendi posse videbantur, quae fortuita sunt atque incerta, tantò admirabilior ars censebitur, cui ista quoque subjacent". Cf. Frans van Schooten, *Exercitationes mathematicae*, Leiden, 1657, p. 519.
[10] Note that the Ars cogitandi and the Logic of PR are the same.
[11] Antoine Arnauld, *The Art of Thinking* (transl. by J. Dickoff and P. James), Indianapolis–N.Y.–Kansas City 1964, p. 354.
[12] Cf. *The Art of Thinking*, p. 355.
[13] Cf. *Leibniz' Sämtliche Schriften und Briefe* (= *LSSB*), ser. II, vol. I, p. 72.
[14] *LSSB*, ser. II, vol. I, p. 489.
[15] Cf. *Nouveaux essais* (French-German edition, ed. and tr. by Wolf von Engelhardt and Hans Heinz Holz). Frankfurt am Main, 1961, vol. II, pp. 268–273, where Leibniz draws attention to his controversy with the sceptic Foucher.
[16] Leibniz, *Nouveaux essais*, vol. II, pp. 508f.
[17] *Nouveaux essais*, vol. II, pp. 508–511.
[18] *Nouveaux essais*, vol. II, p. 512.
[19] *Nouveaux essais*, vol. II, p. 514.
[20] Cf. Leibniz, *Philosophische Schriften*, ed. C.J. Gerhardt, Berlin, 1890, vol. VII, p. 521.
[21] Cf. *LSSB*, ser. VI, vol. I, p. 140.
[22] *LSSB*, ser. VI, vol. I, p. 426.
[23] Leibniz, *Philosophische Schriften*, vol. III, p. 570.
[24] *Werke von Jakob Bernoulli*, vol. III, Basel, 1975, pp. 21–89.
[25] "... id quod valet $\frac{1}{5}$ certitudinis mortis primae seu unam probabilitatem, quarum 5 faciunt omnimodam certitudinem". Cf. *Werke von Jakob Bernoulli*, vol. III, p. 43.
[26] See John Graunt, *Natural and Political Observations mentioned in a following Index, and made upon the Bills of Mortality...*, London, 1662 and frequently thereafter; and Jan de Witt, *Waerdye van Lyf-Renten naer proportie van Los-Renten*, The Hague, 1671. It remains surprising that Bernoulli, who from 1703 to his death in 1705 continually pressed Leibniz to send him a copy of de Witt's work, nowhere mentions the ultimately much more illuminating publication of Halley, which should have been easily accessible to him. See Edmund Halley, 'An estimate of the degrees of the mortality of mankind...', *Philosophical Transactions*, No. 196, 1693, pp. 596–610 and No. 198, pp. 654–656.
[27] Quod et in genere de civilibus et moralibus intelligendum, ubi plerunque unum altero probabilius, satius aut consultius quidem esse novimus, at quot gradibus probabilitatis aut bonitatis antecellet, probabiliter tantum, non accuratè determinamus. Tutissima

probabilitates aestimandi via in istis est non à priori, seu causâ, sed à posteriori seu ab eventu in similibus exemplis multoties observato. Quemadmodum in nostro exemplo, si plurimorum annorum decursu observatum fuisset, duplò plures senes quam juvenculas, ejusdem et aetatis et temperamenti cum nostris juvenculis et senibus, mortuos esse; concluderemus, unum esse casum, qui juvenculae, et duos, qui seni mortem minantur. vel si observatum fuisset, in plurimis senum juvenumque binariis, duplò pluribus vicibus juvenem suo respectivè seni supervixisse, quam senem juveni, concluderemus contrà unum casum, qui juvenem ante senem vitâ privat, duos esse, qui contrarium efficiunt. *Werke von Jakob Bernoulli*, vol. III, pp. 46f.

[28] Quod enim in sortilegis et ludis, quos sola gubernat sors, expectatio praecisè et scientificè determinari possit, causa est, quia accuratè et clarè percipimus numerum casuum, ad quos infallibiliter sequi debet lucrum aut damnum, et quod hi casus indifferenter se habeant et aequè facilè evenire possint, aut saltèm si unus altero sit probabilior, scientificè definire possumus quanto sit probabilior. At quis mortalium obsecro numerum casuum, morborum sc. aliorumve accidentium, quibus obnoxii tum senes tum juvenes, dinumeret, sciatque illos infallibiliter excipi à morte nec ne, determinetque quantò quis altero probabilius grassari possit, cùm haec omnia dependeant à causis omninò occultis et à cognitione nostrâ remotis. *Ibid.*, p. 46.

[29] ***Minùs à verá proportione aberrare possum, si saepiùs quam si variùs observem.*** Ex. gr. Si tot sint casûs pro me, quot contrà me, prodire debet ratio aequalitatis in numero ludorum quos ego & quos adversarius vincit, ponatur ratio quae à ratione aequalitatis parum differat, putà 2 ad 1, dico minorem esse probabilitatem ut ludorum ratio ab aequalitatis ratione magis recedat, cùm 6 ludos ludimus quam cùm tres, & adhuc minorem cùm 9 aut 12 etc. & contrà majorem esse probabilitatem ut intrà tres terminos contineatur cùm plures ludimus quam cùm pauciores. En calculum:

Cum luduntur ludi 3, ut vincam ludos	3, 2, 1, 0
casûs sunt	1, 3, 3, 1 = summa casuum 8
6. . . . . ludos	6, 5, 4, 3, 2, 1, 0
casûs	1, 6, 15, 20, 15, 6, 1 = 64
9. . . . . ludos	9, 8, 7, 6, 5, 4, 3, 2, 1, 0
	1, 9, 36, 84, 126, 126, 84, 36, 9, 1 = 512
12. . . . ludos	12, 11, 10, 9, 8, 7, 6, 5, 4, 3, 2, 1, 0
casûs	1, 12, 66, 220, 495, 792, 924, 792, 495, 220, 66, 12, 1 = 4096

Quare ut plus quam 2/3 partes aut minùs quam 1/3 partem omnium ludorum vincam, cum luduntur ludi 3 probabilitates sunt 1/8

6. . . . . 7/64
9. . . . . 46/512
12. . . . . 299/4096

Est verò $1/8 > 7/64 > 46/512 > 299/4096$ quare constat &c. Imò quod utile & jucundum observare est, fit ut augendo numerum ludorum perveniatur ad probabilitatem datá quâlibet minorem, quoniam ratio singularum ad proximè sequentem maior est ratione 8 ad 7. adeò ut toties observare possim, ut ferè aequè certò probabilitate concludum ac si illas à priori haberem. Quaestio. quot observationes faciendae, ut 100000 mal probabilius sit, ut ratio ludorum, quos vinco, ad illos, quos perdo, magis accedat ad rationem aequalitatis, quam 101 ad 99. *Werke von Jakob Bernoulli*, vol. III, pp. 75f.

[30] Possibile est, tot observationes instituere, ut datá quâvis probabilitate probabilius sit, ut numeri ludorum ab utroque victorum intrà datos limites quantumcunque arctos cadant, quàm extrà illos. *Ibid.*, p. 76.

[31] Contingens (tam *liberum*, quod ab arbitrio creaturae rationalis: quam *fortuitum* et *casuale*, quod à casu vel fortuna dependet) est id, quod posset non esse, fore aut fuisse;... nec enim contingentia semper omnem necessitatem, etiam quoad causas secundas, excludit; quod exemplis declaro. *Ars conjectandi*, Basel, 1713, p. 212.

LUBOŠ NOVÝ

SOME REMARKS ON THE CALCULUS OF PROBABILITY IN THE EIGHTEENTH CENTURY

In to-day's historiography of science an increased interest can be observed in studying the character and differences among the individual phases of the evolution of scientific knowledge. Moreover, it is interesting to note that discussions on a general level, associated, e.g., with the question of scientific revolutions, are conducted systematically and that they have excited interest and a wider response also outside the group of historians of science. On the other hand, the scientific analysis and discussion of these phases in the evolution of the individual scientific disciplines and problems has developed relatively little. I am of the opinion that this fact has several reasons. First of all, a discussion on a general level is easier, because it can draw on various examples from the abundance of the historical evolution of science without having to consider this historical evolution as a whole. It need not, therefore, nor does it solve a number of actual research problems. Another important reason why so far a more profound knowledge of the individual phases, and thus also the transitions between them (revolutions) has not come into its own in studying the evolution of the individual disciplines is the fact that the elaboration of the methodological tools for determining and proving the character and time determination of the phases has only recently become an object of interest. This is associated with the difficulties of the complex analysis of evolution of science in its complicated circumstances, as well as with the terminological apparatus of the history of science being little elaborated. However, I am convinced that a visible, though slow progress is being made in these problems, which should in future become of increasing significance. To the latter statement I have been prompted by the advancement of the history of science itself, as well as by the actual practical requirements to gain more profound knowledge of the regularities in the evolution of science, requirements which stimulate work in the field indeed towards generality and conclusiveness simultaneously.

I have taken the liberty of expressing my general ideas in introduc-

J. Hintikka, D. Gruender, and E. Agazzi (eds.), Pisa Conference Proceedings, Vol. II, 25–32.

ing my commentary also to enable me to draw brief attention to what I consider to be the general features and problems of the topic of the Probability Theory and Probabilistic Thinking in the Seventeenth and Eighteenth Centuries. It is my opinion that this topic, which relatively speaking, has been neglected particularly in the literature pertaining to the history of mathematics[1], has its methodological importance for the history of the mathematical sciences and the history of sciences in general.

In studying the history of sciences in general and the history of mathematics in particular, the researchers display a natural tendency to draw on present ideas and concepts. The naturalness of this tendency is not only implied by the present state being the climax of the whole past development, but particularly the historian, realizing that no state of science is final and that it will be overcome by future development, is indeed forced to utilize to-day's current and predominant system of terms. As regards the theory of probability, which in spite of various difficulties is mostly understood to be a mathematical discipline, it follows that it is understood as such by the historian of science even in the past. There is no doubt that there are sufficient arguments for this understanding; however, it in fact segregates the given problems from their historical context[2]. This slightly obscures the real status of studying probability problems with respect to mathematical sciences, or their parts in the 17th and 18th centuries.

Briefly speaking, the mathematical sciences in the 17th and 18th centuries represented a set of theoretical and practical disciplines, the central concept of which, to a certain extent, were the quantities and the main methodical tool pure mathematics (arithmetic and geometry, possibly algebra and analysis). Although this set of mathematical sciences was very heterogeneous[3], it did display its practical orientation; physical sciences and their applications gradually began to dominate topically, particularly as they ensued from the parallel development of Newton's mechanics. Since the problem of observational errors had still been neglected in the whole of the 18th century, probabilistic deliberations played hardly any part in interpreting mathematical sciences. If we look through the large and widely distributed compendia (textbooks) of mathematical sciences of the 2nd half of the 18th century, we will find sufficient evidence of this. For example, Bézout's[4] course, consisting of several volumes, makes no mention of probability, Karsten's[5] 8 volumes devote a mere

12 pages of the total of 5295 to problems belonging to the probability calculus[6]. The same can be said of Kästner's extensive textbook[7].

It is true that further examples, which would not bring such unique evidence at first glance, could be given. However, it should be pointed out that the association between the probability calculus and combination calculus was very vague in the 17th and 18th centuries[8]. Stimuli for deliberations in combination calculus came from other fields (e.g., the powers of polynomials, recreational mathematics) and even Hindenburg's school of combination calculus, which developed in the 2nd half of the 18th century, not only drew on different concepts, but also claimed different purposes for this calculus (among others also a tool of infinitessimal calculus).

These examples allow us to formulate the hypothesis that the study of the probability calculus in the 18th century developed mostly in close association with heterogeneous problems, for the accurate and mathematical solution of which it created a common base[9]. The literature on the development of the probability calculus as well as the papers, prepared for this meeting, indicate a sufficiently large variety of different regions.

In order to illustrate the position of the probability calculus in the 2nd half of the 18th century, I should like to draw attention to one important work which, perhaps in a slightly distorted manner, reflects the scale of fields, in which probabilistic thinking was utilized, in a broad context. The work I refer to is the French Encyclopaedia[10]. It contains no direct keywords referring to the probability calculus or its parts and various cases are frequently scattered in a very special form under various names. Nor is the probability calculus mentioned in the system of sciences, but only "arithmétique politique" is given[11]. In the keywords themselves the problems of the probability calculus are concentrated under keywords referring to gambling, to probability (*Probabilité*[12]) and around questions which may perhaps be expressed by the term "arithmétique politique", then used, which is perhaps most extensive including its practical applications[13]. Without going into details, we shall restrict ourselves to giving some of the general features of these keywords.

The central keyword is "Probabilité", which understandably, does not belong just to the field of mathematics, but also to philosophy and logics. In the introduction probability is explained by our partial knowledge which prevents us from having absolute certainty about the truthfulness of one or another statement, but only probability "dont une juste estime et une exacte mesure seraient le comble de la sagacité et de la prudence". The keyword then adds: "Les Géomètres ont jugé que leur calcul pouvait servir à évaluer ces degrès de *probabilité*, du moins jusqu' à un certain point, et ils ont eu recours à la Logique, ou à l'art de raisonner, pour en decouvrir les principles, et en établir la théorie". After a brief explanation of the foundations of a probability calculation[14], it is shown, how "wise and careful people" make use of probabilistic judgement in practical life[15]. Although a discussion of the source of probability follows, which would enable also the mathematical aspect to be developed[16], more attention is devoted to practical advice and conclusions. It is particularly emphasized that "l'expérience du passé est un principe de *probabilité* pour l'avenir; que nous avons lieu d'attendre avec raison des événements conformes à ceux que nous avons vu arriver; et que plus nous les avons vu arriver fréquemment et plus nous avons lieu de les attendre de nouveau. Ce principe reçu, on sent de quelle utilité serait dans les questions de Physique, de Politique, et même dans ce qui regarde la vie commune, des tables exactes qui fixiraient sur une longue suite d'événements la proportion de ceux qui arrivent d'une certaine façone à ceux qui arrivent autrement...". In this connection the keyword also emphasizes personal qualities: "L'attention donnée au passé, la fidelité de la mémoire à retenir ce qui est arrivé, l'exactitude des régistres à conserver les événements, font ce qu'on appelle dans le monde l'expérience. Un homme qui a de l'expérience est celui qui ayant beaucoup vu et beaucoup réfléchi ..." may estimate the probability.

The extensive text gives examples and shows the necessity of analysis for evaluating the probability of events. For example, it is required that one first investigate the possibility of existence of a phenomenon before determining the probability of its existence on the basis of evidence (of reliable and less reliable witnesses).

After drawing attention to literature and associated keywords[17], the conclusion rings optimistically: "Conclusons qu'il ne serait pas entièrement impossible de réduire toute cette théorie des *probabilités*[18] à un calcul assez reglé, si de bons génies voulaient concourir par des recherches, des observations, une étude suivie, et une analyse de coeurs et de l'esprit, fondés sur l'expérience, à cultiver cette branche si importante de nos connaissances, et si utile dans la pratique continuelle de la vie. Nous convenons qu'il y a encore beaucoup à faire, mais la considération de ce qui manque doit exciter à remplir ces vides, et l'importance de l'objet offre de quoi dédommager amplement des difficultés".

The other related keywords do not deviate from the direction we have indicated and they are all founded on the keyword *Probabilité*[19]. Let us consider three of these: *Certitude, Hazard* and *Hypothése*[20]. In the first of them, *évidence,* or as d'Alembert wrote in his Discours préliminaire, *l'évidence de la certitude,* which is inherent to mathematics, is distinguished from "certainty", which is mediated and which occurs in all other fields, including physics, it being particularly stressed that it is necessary to distinguish between the probability and certainty of a statement, moral certainty being involved frequently. The keyword Hazard strongly emphasizes that the randomness of phenomena does not follow from them having "une cause nécessaire ou prévûe", but that we only do not know the causes which with these phenomena are the same "nécessaires et déterminées" as with others. The concept as well as scientific meaning of "hypothesis" is interpreted in the same spirit: "Les véritables causes des effects naturels et des phénomènes que nous observons, sont souvent si éloignées des principes sur lesquels nous pouvons nous appuyer, et des expériences que nous pouvons faire, qu'on est obligé de se contenter de raisons probables pour les expliquer. Les Probabilités ne sont donc à rejetter dans les sciences . . .". The subsequent text then defends the significance of hypotheses in science, which are based on partial knowledge, emphasizes the necessity of forming them, although they are only probable and may be erroneous. This keyword, like many others, contain no indication of a possible calculus.

The reflection of the probabilistic thinking and probability calculus in the French *Encyclopédie* indicates that even in the 2nd half of the 18th century there existed a wide field in which probabilistic ideas, although as yet vague, asserted themselves and that they were of a profound practical importance; this region may be expressed, to a considerable extent, by the term "low sciences". The solving of their problems by means of precise methods of mathematical, quantitative calculus was the aim, however, it is hard to evaluate the difficulties to-day which this effort encountered theoretically (establishing of accurate mathematical tools), as well as regards practical applications which remained restricted to a small section of relatively simple cases. It is my opinion, therefore, that in the 2nd half of the 18th century there still existed a close symbiosis of probabilistic thinking and developing probability calculus and that they cannot be divided precisely. One may perhaps say – and this may also be true of periods other then the 18th century – that, from the point of view of mathematics of the time, the probability calculus represented one of the few cases of the forming of a mathematical discipline in close contact with broad and heterogeneous practical fields, in which probabilistic thinking was being effected.

In this connection I should like to draw attention to one problem which is sometimes neglected in the history of sciences. Science should, one may perhaps say truthfully and conclusively, solve problems of a given field so that the practical exploitation of its knowledge is reliable. The question of the accuracy of even the fundamental terms (and of the terminological system) is frequently pushed into the background; the development of infinitessimal calculus in the 18th century is proof of this. This was also the case in developing the probability calculus of the 18th century, or was the question of clearly defined terms in a special form so important, that it directly limited the extent of applications of the mathematical theory?

To conclude I should just marginally like to point out the way in which Bernard Bolzano introduced the term probability. In his treatise Wissenschaftslehre[21] he arrived at the term probability by studying the relation between propositions (*Sätze*)[22]. His procedure was as follows[23]: Already earlier he had introduced the concept of a proposition with variable notions. We shall denote it A(i, j, . . .). We shall restrict ourselves to the set of these notions for which the proposition has a meaning. The set of notions, for which proposition A is true, is the subset of the original set[24]. Now, consider propositions A, B, C, D, . . . with variable notions i, j, . . .; "so wird es öfters ungemein wichtig, das Verhältniss zu erfahren, in welchem die Menge der Fälle, darin die Sätze A, B, C, D, . . . alle wahr werden, zur Menge derjenigen Fällen stehet, in welchem neben ihnen auch noch ein anderer Satz M wahr wird". The probability of proposition M in dependence on proposition A, B, C, D, . . . and notions i, j, . . . is then the ratio, "in welchem die Menge der Fälle, worin A, B, C, D, . . . wahr werden, zur Menge derjenigen stehet, wo neben ihnen noch M wahr wird, ob wir auch M für wahr annehmen sollen oder nicht". Bolzano then goes on to show that, on the basis of the definition, he is able to derive not only a proposition on which the probability calculus is founded, but also solve ordinary practical problems[25].

NOTES

[1] I do not want to claim that historical literature on the theory of probability is less extensive than that concerning other disciplines. However, the attention devoted to its relation to probabilistic thinking, on the one hand, and to integrating it into the mathematical development as a whole, on the other, is small; for example, the recently published review by Morris Kline, *Mathematical Thought from Ancient to Modern Times* (New York 1972) does not devote any attention to the theory of probability and, according to the Subject Index, the term "probability" only occurs once.

[2] This feature comes into the limelight particularly in those historical treatise which attempt a more comprehensive (or more systematic) coverage of a longer period of development of the study of probabilistic problems. This also includes an older paper by T. Todhunter, *History of the Mathematical Theory of Probability* (Cambridge-London 1865), as well as Л. Е. Майстров, Теория вероятностей, Исторический очерк (Москва, 1967, 318 pp.). In French literature of the 18th century the term "calcul des probabilités" is more widespread.

[3] For details refer to L. Nový, 'K pojetí vědy na počátku průmyslové revoluce (On the conception of science in the beginning of Industrial Revolution)' *Práce z dějin přírodních věd* 6, Praha 1978, p. 123–135.

[4] E. Bézout, *Cours de mathématiques . . .*, Paris 1770.

[5] W. J. G. Karsten, *Lehrbegrif der gesamten Mathematik*, Theil I–VIII, Greifswald 1767–1777.

[6] Teil II. Der X. Abschnitt: Von der Annuitären– und Leibrenten–Rechnung, p. 282–293.

[7] A. G. Kästner, *Anfangsgründe der Mathematik*, Erste Auflage. Göttingen 1758 n, Sechste Auflage, Göttingen 1800.

[8] Treatises on the history of mathematics in 17th century summarize these regions with time; compare R. Taton (ed.) *Histoire Générale des sciences*, Tome II, p. 219, or A. П. Юшкевич, История математики, Том второй, p. 81–97.

[9] In no way do I wish to throw doubt on the existence of the relatively independent scientific field of "probability calculus"; I just wish to avoid this question.

[10] *Encyclopédie ou Dictionnaire raisonné des sciences, des arts et des métiers*, Vol. I–XVII, 1751–1765. The mathematical parts could have been influenced by d'Alembert even after the VIIth volume had been published; his relation to many of the results of the probability calculus was sceptical. Compare, e.g., Майстров, p. 122, 145 ff, who also gives the possible reasons for d'Alembert's erroneous probability calculations.

[11] Tome I. I have avoided here discussing the associations and meaning of this system of sciences, reproduced in the introduction.

[12] Tome XIII, p. 393–400.

[13] The keyword "arithmétique politique" (Tome I, p. 678–680) in itself rather explains the practical meaning of accurate statistical data on society and of the calculations which can be made on its basis even for estimating future conditions. The individual

questions are then developed by other keywords, e.g., the keyword *Vie* (*durée de la vie*), which belong, as mentioned, to political arithmetic (Tome XVII, p. 249–254), as well as the complex of keywords referring to dues, insurance, etc.

[14] It is surprising that no mathematical apparatus is used here, although the Encyclopédie contains many keywords relatively exacting mathematically.

[15] In the rules it is said, among other things, that it is difficult to determine accurately the degree of probability of statements and that only genii can do so and that they were rare (Rule 4).

[16] Two sources of probability are distinguished: the first comes from the knowldege of the nature of things and the second from experience, when we draw on the past to estimate the future.

[17] He refers to treatise by Deparcieu, Montmort, Jacq. Bernoulli and Halley. Of related keywords he points out in particular Induction, Analogie. Hypothèse, Supposition and Combinaison, which are all outside the sphere of the mathematical theory of probability.

[18] This formulation which is characteristic of understanding probabilistic problems in the French Encyclopédie, emphasizes the difference between the theory of probability, i.e. a region which, on the basis of various problems and examples mostly of a practical significance, constitutes a theory of the probability region, and its "calculational" elaboration, which could possibly be called "calcul des probabilités".

[19] Other keywords, in their aspect of content, are directly founded on more recent literature, or they contain their analysis and discussion. I should say that considerable weight was being attributed to statistical data, references to them, and practical calculations, following from them.

[20] 'Certitude', Tome II, p. 845–862: 'Hazard', Tome VIII, 'Hypothèse', Tome VIII, p. 417–418.

[21] B. Bolzano, Wissenschaftslehre. Versuch ausführlichen und grösstentheils neuen Darstellung der Logik mit steter Rücksicht auf deren bisherige Bearbeiter. Band I–IV, Sulzbach 1837.

[22] Band II, §161, str. 171–191; §167, str. 206; §253, str. 509–514.

[23] We are attempting to express the essence of Bolzano's procedure without distortion and excessive modernization.

[24] Bolzano currently used the term Menge here, he did not introduce the term subset, which he expresses periphrasically, and confines symbols to a minimum.

[25] In the spirit of his philosophical approach he expresses himself in favour of an objective understanding of probability; in his formulation this means that the given relations also hold for "Sätze an sich" (compare p. 185 f.) As usual in his comment (p. 187 ff) Bolzano refers to several authors whose opinions he discusses. These include Aristotle, Cicero, Locke, Huyghens, Wolf, J. Bonnet, Lacroix and Laplace.

V.N. KOSTIOUK

PROBABILITY AND THE PROBLEM OF INDUCTION

In Prof. Hacking's monograph *The Emergence of Probability* an interesting outline of the genesis of the concept of probability was presented. I should like to discuss a problem mentioned by Hacking. It is Hume's skeptical problem of induction.

This problem is characterized by the following questions. What are the historical-scientific conditions of the appearance of induction problem formulated by Hume? Why was this work of Hume's published in 1739 and not a century sooner or later.[2] What is the relation of this problem to the concept of probability? To Berkeley's philosophy? Why wasn't this problem formulated by the universal genius of Leibniz who was interested in induction and was in a certain sense the founder of inductive logic?

In order to answer these questions, it is necessary, first of all, to analyse the concepts "knowledge" and "probability" in their historical context. According to Leibniz, knowledge is proof based on the law of sufficient reason.

Such understanding of knowledge was in approximate conformity to a historical tradition which goes back to Aristotle. According to this tradition, knowledge is a proof from some reliable first principles (deductive ideal of knowledge as simplified by the scholastics).

It is obvious that such understanding of knowledge (such ideal of knowledge) excludes a possibility of formulating the problem of induction) as it is interpreted by Hume, because the latter proposes a broader conception of knowledge. In this conception the results of observations play a more important role.

Leibniz' interest in inductive logic was in conformity with the deductive ideal of knowledge. His interest in inductive logic is but an interest in the (analytical theory of probabilities. According to Leibniz, the science of probability should become "a new sort of logic" in which conditional probability is understood as a generalisation of deductive inference (the inference in which conclusion follows from premises with maximal probability).

As it is well known, this idea was developed further in the 20th

J. Hintikka, D. Gruender, and E. Agazzi (eds.), Pisa Conference Proceedings, Vol. II, 33–40.

century (for example, by R. Carnap). Leibniz knew already that a probabilistic sentence may express a relation between hypothesis and evidence. But he did not attach a broader significance to this relation which would go beyond the boundaries of conditional probability.

The development of early modern science was a circumstance that played an important role when the problem of induction came into being. It is associated with the works of Copernicus, Galileo, Newton and others, and with the creation of classical mechanics. As a result of all this, the scholastic conception of knowledge disappeared, and two relatively independent forms of scientific knowledge emerged: mathematical knowledge and natural-scientific knowledge. From the point of view of our problem we are interested only in the latter.

But here a question arose what the laws of nature are and, in particular, what the cause-and-effect relations are. Many philosophical systems of the 17th and 18th centuries tried to answer this question. Berkeley, in particular, wrote: "The connection of ideas does not imply the relation of cause and effect but only a mark or sign with the thing signified". With this in mind, it is sometimes said: Hume becomes possible (after Berkeley).

This point of view is accepted by Hacking, and it leads him to a more or less usual reconstruction of Hume's views. According to Hacking, the transformation of the concept of cause and effect from knowledge to opinion which, unlike knowledge, is not certain but only probable served as the direct stimulus of the skeptical induction problem of Hume. The result of this is that an interpretation of causality as the simple regularity of events in time, as habit and custom, becomes possible, and this opens the way for the sceptical problem of induction in Hume's usual formulation or in one close to it.

It seems to us that such an approach to the genesis of induction problem is not quite correct or is not unique. This approach implies a much too high degree of closeness between the whole of Hume's philosophy and the problem of induction as such which, as we see it, is mostly independent of the remaining part of Hume's philosophy (it depends much more on the development of science in the 17th and 18th centuries). And if one looks for the philosophical prerequisites that make possible the appearance of the (sceptical) problem of induction, it is better to turn one's eyes on Locke rather than on Berkeley.

When the the genesis of the sceptical problem of induction is reconstructed, one can choose a way different from that of Hacking's. One has to answer the question: what is the objective meaning of the sceptical problem of induction (which, naturally, may not coincide with the subjective opinion of Hume himself)?

As a matter of fact, Hume's problem of induction is, a part from not very significant details, a discussion of the justification of non-mathematical knowledge (i.e., both the ordinary and the scientific one). The fact that this problem was lent a paradoxical character by Hume himself is not very essential.

As it seems to us, there is a certain analogy between the philosophy of Eleatics and the sceptical problems of induction in Hume's formulation. The Eleatics apparently wanted to demonstrate that motion is impossible, but as a matter of fact they discussed certain difficulties associated with the understanding of motion. Similarly, Hume apparently wanted to demonstrate by his sceptical problem of induction the fundamental uncertainty of non-mathematical knowledge based on the concepts of cause and effect, but he essentially showed that the empirical view on knowledge (originating from Locke) could not serve, in combination with the deductive ideal of non-mathematical knowledge, as a reliable foundation for its justification.

Under the conditions of such an approach one discovers that the sceptical problem of induction is independent of the concept of probability (although there is a certain connection between them). For the (sceptical) problem of induction to appear, the following three conditions are necessary (and maybe sufficient together).

(1) The appearance of the experimental science of the early modern period.

(2) The idea that it is impossible to eliminate the hypothetical element from scientific knowledge. Roughly speaking, scientific knowledge can be presented as a combination of hypotheses that correspond to empirical evidence more or less fully but are not deduced from them this evidence (and therefore need justification).

(3) Only the deductive proof or deductive inference can serve as such justification (according to the deductive ideal of knowledge). Hypotheses are not deduced from evidence that is why one must find an additional principle (like a principle of sufficient reason) which will allow us to construct a deductive inference leading to some of these hypotheses.

Note that condition (2) is much more general than just a statement of the relation between the two components of the relation of conditional probability. As a matter of fact this condition does not depend at all on the concept of probability, but instead depends on the general conception of sensualism that goes back to Locke.

Thus the sceptical problem of induction depends neither on the concept of probability nor on Berkeley's philosophy. To put it more precisely, Hume's philosophy on the whole shows such dependence and the induction problem itself does not.

Note also that there is a faint agreement between the second and the third condition. It is exactly this which leads to the paradoxical character of the induction problem in Hume's formulation. But such peculiarity of this problem is not inevitable. A possible (non-sceptical) solution of this problem may be associated with the rejection of third condition and the transition from the deductive to the inductive ideal of knowledge, which emphasize more conjecture than logical inference.

The result of this is that the vicious circle mentioned by Hume disappears, and the problem itself is formulated in other concepts. Therefore, this will be a different problem. To distinguish the latter and the sceptical problem of induction in Hume's formulation, we shall call this new problem the "problem of induction" omitting the term "sceptical".

The formulation of Hume's sceptical problem of induction is simple in a certain sense and can be reduced to the following argument. Only necessary statements can be proved logically. But the statement "such-and-such cause at a moment T_1 produced such-and-such effect at the moment T_2" is not necessary and therefore cannot be logically proved. A possibility remains that this statement can be deduced from an empirical postulate. This postulate, however, whatever its concrete formulation is, is not, if it exists, necessary, and for this reason cannot be proved logically. Its proof, if it is possible, requires another empirical postulate, and so on to infinity. It is possible to formulate this reasoning in a different way when, instead of the explicit regress into infinity, an explicit vicious circle appears.

Apparently, one of the ways out of this paradoxical situation noticed by Hume consist in the rejection of the deductive proof (and deductive inference) as a means of justification of non-mathematical knowledge (the rejection of the deductive ideal for such knowledge).

As a result the necessity of finding an additional postulate disappears. One has to recognise honestly that such knowledge contains hypotheses subject to checking for which there are no deductive proofs from "principles". As a result of this, a new problem appears, that justification some of these hypotheses which is not reducible to deductive inference (from unknown premices). It is exactly what constitutes one of the problems of induction. This problem contains nothing sceptical or agnostic.

Such a problem of induction can be called the problem of inductive justification of non-mathematical knowledge. Its essential feature is the rejection of proof as the main means of justifying knowledge (proof and especially deductive inference can nevertheless be an important auxiliary means of such justification).

This is the principal changing because it is a conceptual one. Two new inductive concepts: (a) confirmation of hypothesis by evidence, (b) acceptance of hypothesis on the basis of evidence, occupy now the place of deductive proof.

These inductive concepts differ essentially from the conditional probability (much has been to support this view done by K. Popper) but they can be described, as a rough approximation, by means of certain functions of probabilities or by means of certain verbal rules whose formulation contains a reference to probability. In principle, however, the reference to probability in their definition is not necessary. Note also that there is a substantial difference between these two inductive concepts.

In order to describe the difference between them, let us first describe the presence of the two aspects of the concept of probability: the objective and the subjective. One of the main merits of Hacking's book, *The Emergence of Probability*, is the clear observation of this distinction from the historical-scientific point of view. Without such distinction it is impossible to understand correctly the history of the theory of probability.

It is necessary, however, to complement the usual distinction between the objective and the subjective aspects of the concept of probability. In addition to the equivalent objective and subjective interpretations of the calculus of probabilities, there is a more general philosophical concept of probability which is based on the subjective approach but is, nevertheless, quite objective. This concept expresses the fundamental fact that the human being in his activity makes estimates of probabilities that are subjective as human estimates are.

As a rule, these estimates contain errors and differ from the objective probabilities. Fortunately, the human being possesses the ability to correct these estimates as he obtains new knowledge (by making use of, for instance, Bayes's theorem). The results of this procedure are sequences of such estimates.

In many cases these sequences of probability estimates used by a human being converge to their true (objective) values (if they exist). In this sense the subjective conception of probability is quite objective. The question of whether such convergence always takes place, however, remains open (the perspectives of knowledge always constitute an open problem).

Further, the difference between the objective and the subjective aspects is essential not only for probability but also for the inductive concepts of justification of knowledge (as it is, incidentally, for many other concepts). The confirmation of a hypothesis by evidence is entirely objective; it depends just on the relation between hypothesis and evidence (attention to this was paid by Carnap) and does not depend on the human being who uses these hypotheses and evidence nor on the nature of human knowledge. But the acceptance of a hypothesis on the basis of evidence or its rejection is always done by a human being. It is subjective in this sense: it depends on the level of knowledge, on one's epistemic values.

The acceptance and the rejection of a hypothesis depend not only on the available evidence (as in the case of confirmation) but on other factors as well. Even in the presence of the same evidence different people may accept different hypotheses (from the same list of hypotheses) since their estimate of this evidence and of the hypotheses is different. Therefore the acceptance or rejection of a hypothesis is not a purely logical matter but rather a historical-scientific relation which includes a reference to the subject of knowledge and to the historical nature human activity.

(Note in parentheses that for modalities it would be also useful to observe the similar distinction. For instance, necessity is quite objective, whereas various types of epistemic modalities: know, believe, assume, etc, must include a reference to the subject of knowledge or cognition. Obviously, this does not deprive epistemic modalities of their objective content.)

There are also other differences between confirmation and acceptance. Let $h_1, \ldots, h_n$ be an exhaustive list of alternative hypotheses, and e the evidence used. Then this evidence may confirm $1 \leq k \leq n-1$ hypotheses from this list, but only one of them can be accepted (by the same person). Evidence corroborating different alternatives cannot always be significant enough to enable one to choose between them. (This explains the fundamental fact that for the confirmation of a scientific hypothesis it is not sufficient for it to be accepted.)

Transition to an inductive ideal of scientific knowledge is not confined to the replacement of deductive proof by inductive acceptance. More significant is the recognition of a certain intellectual climate directing the course of theoretical modfications in each branch of scientific research.

The necessity of taking into account all proposed hypotheses (including those which are not accepted at present) has at least three grounds.

1. None of these hypotheses, not even the least probable and most extravagent one, is deductively refuted; that is why it might be true.

2. The degree of confirmation of these hypotheses can change in the course of time.

3. Notwithstanding the fact that most of these hypotheses are considered to be false, they are not as useless as they seem to be, as far as the alteration of knowledge is concerned: the way of treating them as false influences the quest for hypotheses with a broader perspective and promotes the discovery of such hypotheses. Thus the rejection of hypotheses is a continuous process, in which the non-acceptance of one hypothesis has an effect on the formation of new hypotheses.

At last, I must draw the following in conclusion. Prof. Hacking has made a very interesting objection. He asks: Maybe this point of view is a return to the analytical problem of induction which was known in 1660, not in 1739.

Maybe he is quite right, but I have some doubts about it. The problem of induction which is analysed by myself exhibits certain conceptual differences from the analytical problem of induction according to Hacking. The second problem is the problem of analysing non-deductive "inference" by means of the concept of probability. It is a logical rather than epistemic problem. The first problem is the problem of justification of non-mathematical knowledge, which is not in principle dependent on probability concepts and is not purely logical.

According to Hacking, for the transition to the inductive problem one needs a change in our understanding of the concept of cause. In my view, a transition to a new (non-deductive) ideal of knowledge is necessary. Such a problem could not be posed in 1660. The step taken by Hume in 1739 is but a half-way station in this direction. A replacement of the ideal of knowledge (which means the final reduction of the problem proposed by Hume) was left to the 20th century to carry out.

ANNE M. FAGOT

PROBABILITIES AND CAUSES: ON LIFE TABLES, CAUSES OF DEATH, AND ETIOLOGICAL DIAGNOSES

Contents

... ici, comme dans le reste de la vie, la confiance diminue à mesure que les lumières augmentent...
Condorcet, *Oeuvres*, t.I, p. XIII.

J. Hintikka, D. Gruender, and E. Agazzi (eds.), Pisa Conference Proceedings, Vol. II, 41–104.

ANNE M. FAGOT

PRELUDE: THE CAUSE OF DEATH STATISTICS HISTORICAL BACKGROUND

Causes of death began to be recorded by London parishes as early as 1592: "*the occasion of keeping an accompt of burials arose first from the plague*", said John Graunt [1662, §1]. From 1629 on, the London Bills of Mortality gave detailed weekly returns of diseases and casualties enumerated in alphabetical order (lists of 60 to 70 causes). The registration of causes of every person's death became mandatory in England under the 1837 Registration Act.

During the first half of the 19th Century, several other European cities attempted to inquire into the causes of death with a view to both medical discovery and improvements in sanitation.

The International Statistical Congress held in Paris in 1855 proposed a statistical nomenclature for the compilation of cause of death returns: simultaneously translated into seven languages, the list of 139 'morbid entities' was a compromise between William Farr's (Registrar General of England and Wales) and Marc d'Espine's (Geneva) projects. It was never used.

When Jacques Bertillon became chief statistician of the 'Bureau de Statistique de la ville de Paris' in 1883 he carefully drew up a nosological nomenclature (along Farr's lines) which was put to use in Paris in 1886.

Bertillon's nomenclature including 161 items was adopted by the International Statistical Institute (Chicago, 1893) and recommended for use by all the registrars of vital statistics in the United States, Canada and Mexico by the American Public Health Association (1897, Philadelphia). It thus became the International Classification of Causes of Death, submitted to decennial revision, and first revised by an international conference held in Paris in 1900 with the view that 20th century statistics should be internationally comparable.

The sixth decennial revision conference (W.H.O., 1948), besides making important changes in the classification (list of about 1000 3-digit categories), introduced an international form of medical certificate of cause of death. This format had been tried in England for many years, and specific recommendations for certifying doctors, as well as instructions for coding (and interpreting) medical diagnoses for statistical tabulation, carry an implicit doctrine about causation of death, which may be considered the outcome and culmination of some two hundred years of international reflections and (oftentimes hot) discussions.

The international certificate of cause of death, and the International Classification of Diseases, injuries and causes of death, are currently in use in all the nations members of the W.H.O. (9th revision effective Jan. 1, 1979).

The sources of the international doctrine of the causation of death may be traced to both the beginnings of anatomo-clinical method on the physicians' part, and early probability theory on the statistical side. This paper goes back to some of these sources in order to find out whether there is a connection between what a doctor and what a statistician call a *cause* of death (other than the trivial fact that the second supposedly compiles whatever the first writes out).

* * *

0.0. During the 1870 siege of Paris, the number of deaths increased dramatically: from an annual average of 45,000 for the previous years, it went up to 78,581 that year, out of which about 30,000 were for the period from September 20 to December 31, coinciding with the siege. The weekly returns published by the city's administration show a +62% surplus mortality in October, and a +70% surplus in November. Historians report that the people of Paris were aware of this fact because of the constant passage of hearses through the streets. As for the medical officers, they "could hardly suffice day after day to their task of certifying deaths".

Doctor Vacher (a physician in charge of ambulances) at the beginning of 1871 conducted a study on the causes of the increase in the mortality rate for the *Gazette Médicale.* His study was reproduced in the *Gazette des hôpitaux,* followed by an analysis of death returns for January 1871, when the city was shelled.

Vacher's study, published in well-circulated medical journals, gives a view of how physicians assimilated the elements of a statistical methodology which had been worked out by mathematicians some decades before.

"One would be tempted" said Vacher, "to credit the siege, or rather its concomitant hardships, with the dramatic rise in death rates: however, when examining the matter more closely, one is soon convinced that *such an excess mortality is the result of a variety of causes, some of which are foreign, nay anterior to the state of siege*" [*Gazette des hôpitaux,* 24 Janvier 1871, page 39].

Vacher then tried to evaluate "*the part of influence*" attributable to each of those causes.

As the enemy approached, the people living in regions adjacent to the capital had flowed in and taken refuge within the city's walls. Their number could be estimated quite accurately by comparing the returns of the 1866 census (around 1,800,000 inhabitants), with those of the October 1870 census, hastily conducted for the purpose of organizing the rationing of meat (over 2 million). If soldiers and mobile guards (the number of which would have been around 260,000) were also counted, a fair estimate was that the city's population had increased by some 29%, i.e., well over a quarter, which by itself entailed a proportional increase in mortality, having nothing to do with the war (mathematically, if not historically).

Hence there was a "*first cause, independent of the siege*", to which some 30% out of the 70% observed surplus mortality could be charged. One was left with a 40% surplus to account for. "*The increase in population*", Vacher went on, "*does not alone explain the increase in mortality,... a complement to the explanation has to be looked for elsewhere*" [*Gazette des hôpitaux*, 1 Juin 1871, p. 259].

It will be noticed here that, just as in the common clinician's vocabulary, *the cause explains the effect, in proportion to the influence it has been found to have in its production.* And the *complement to the explanation* is a *complementary cause.*

0.1. Smallpox ravaged the population during the months of the siege; with about 5000 deaths from October to December, it accounted for practically all the remaining excess mortality (+38% in October, +45% in November). And even though a slight new outbreak could be delineated towards the end of the year, there was evidence that the epidemic "*existed before the siege*" (5000 deaths from January to September, against an annual 'normal' average of about 500).

Parenthetically Vacher called attention to two traits of smallpox mortality which the statistics revealed. First, the old and the newborn *also* died of smallpox, which "refuted the belief" in an endurance or immunity proper to extreme ages, and *expressly* indicated that everyone should be vaccinated. Second, two public hospitals (La Salpêtrière and Lariboisière) stood apart with quite an "excessive" smallpox mortality relative to the others. Vacher looked for a *special cause* of the difference and suggested it might be connected with the treatment of laundry, for the disease's transmission could be followed as it spread from washing places which handled the contaminated linen. He took the opportunity to throw the blame on the incompetence of the directors of the 'Assistance Publique', and claimed "that hospitals should be run by physicians", as in Germany and England.

Smallpox, as an explanation of mortality, was therefore in itself a complex cause, since on top of the recrudescence of the epidemic, there was an administrative slackness, as well as overpopulation (for overpopulated districts were afflicted more than others).

Increase in population together with the smallpox epidemic: "*the combination of those two causes already suffices to explain mortality's aggravation*" [ibid]. Was there a need to go farther?

0.2. The people of Paris complained about "the rigors of a severe winter, worsened by food shortage" [ibid]. One could see frozen soldiers, and citizens struck in the streets by apoplexies *a frigore.* However, a minute examination of cause-of-death returns did not, for example, reveal any increase in the rate of respiratory diseases (it even showed a decrease). So, one might have concluded that, apart from smallpox, the sanitary situation was rather satisfactory, were it not for a (very) few cases of rare diseases that had never been reported in the past. Vacher found *prima facie* explanations (i.e., causes) for them: *scurvy* might have been due to the use of tinned meat instead of fresh meat; *dysentery* could be traced behind the 135th battalion as it moved from Charenton to Ménilmontant; *intermittent fevers* were localized around the St. Martin canal, the waters of which had been diverted upstream by the Prussians, so that putrid smells pervaded the area until autumnal rain washed the stagnant waters away into the Seine.

The question was whether the appearance of (qualitatively) unwonted causes of death (whose quantitative influence in the overall mortality was very feeble) announced a deep "*silent cause of death*" [*Gazette des hôpitaux*, 3 Juin 1871, p. 263–4] connected with the siege.

In order to make up his mind, Vacher went back to the only similar case in history, about which he was able to collect some information: the four-month siege Paris sustained in 1590, during which the city's authorities (the leaguers) organized the rationing of food supplies, just as was done in 1870. Parish registers showed that the siege "took a deadly turn" (244 burials in the St André des Arcs parish, for example, against a yearly mean of 42) when, after consuming all the available edibles including "horses, donkeys and dogs", people started being short of necessities.

"It is important to be specific about the *effects* of insufficient food" [ibid.]. Vacher reflected on his own observations as a doctor. Forced into an almost vegetarian diet (the meat ration had been reduced from 300 grams/day to 30 grams/day, a "palate-revolting" meat-fat was sold for butter), *adults* lost weight and their performance did not seem to be impaired *unless* they fell ill. Indeed, physicians had noticed that most diseases had become "*more serious than in ordinary times*". Typhoid, for example, was lethal in 40% of the cases, instead of the usual 25%. Young *children* however showed all the signs/effects of inanition. "Progressive emaciation, ... oedema, anemia, incoercible diarrhea, continuous plaintive cries, ... are the symptoms characteristic of the starvation fever which currently thins out our infant generation" [ibid.].

Vacher did not conclude that the people of Paris were *dying of hunger* (in fact, death certificates hardly ever mentioned such a cause of death). He concluded that undernutrition might be an *underlying* potential cause which aggravated the seriousness of all conditions, and was likely to amplify their deadly consequences over time; on which hypothesis the physician could predict higher mortality rates from all causes for the immediate future.

Though they much impressed the people's minds, war actions themselves had little bearing on mortality: there were fewer casualties in the various engagements (3000) than from pneumonia (3500), fewer people were killed by shellings (44) than by typhoid fever (300).

0.3. After reducing the notion of *cause* to that of *invariable succession* ("*the invariable antecedent is termed the cause, the invariable consequent, the effect*" [1843, Book III, Ch. V, §2]), John Stuart Mill acknowledged the fact that in the field of medicine (as in others) there are cases in which the same effect can evidently be produced by a variety of causes ("*many causes may produce death*"), and in which the same cause may have different effects according to the way it is counteracted. Hence, the necessity to resort to statistics, when a careful analysis of the cases in which the effect obtains does not allow the isolation of one *constant antecedent*, but merely uncovers a number of *frequent antecedents.* "*The plurality of causes is the only reason why mere number is of any importance*" [1843, Book III, Ch. X, §2]. As to the "mixing of effects", in order to save the principle of causality, Mill was forced into holding that "*all laws of causation, in consequence of their liability to be counteracted, require to be stated in words affirmative of tendencies only, and not of actual results*" [1843, Book III, ch. X, §5]. The same causes do not always produce the same effects, but they always *tend* to . . . ."

Thus, typhoid fever would naturally 'tend' to kill 25 times out of 100, undernutrition would not 'tend' to kill adults, while in fact a typhoid fever would kill 40 out of 100 undernourished adults . . .

Now, the utility of resorting to numbers when one wants to infer causes from observed effects is exemplified by Vacher's inquiry. As far as 'tendencies' are concerned, it is doubtful that Vacher (or any physician) would presume that typhoid fever 'tends to' produce tuphos, diarrhea, myocarditis . . . For, either the 'effect-symptom' is *by definition* a characteristic of the disease, in which case typhoid fever will not be diagnosed in its absence, or it is an inconstant effect, in which case there does not seem to be a need for keeping within the 'essence' of the disease all its possible complications in order to save pseudo-universal causal laws. One can estimate the frequency of the complication and do away with the law: the fact that typhoid fever kills in only 25% of the cases does not seem to prevent one from asserting that it causes death when it does, without having to hold that it is '*potentially always mortal*'.

The point is that a probabilistic approach, together with accurate definitions (not the recourse to universal laws) is quite congenial to medicine, when it comes to inferring causes from observed effects. As to the development of effects from their causes, it is substantiated through the history of the patient's case.

* * *

1.0. *Inoculation* of variolous matter for protection against smallpox was introduced in Europe in the second decade of the 18th century (England, Lady Montagu, 1717). The method is thought to have been imported from China *via* Greece and Turkey. It was supplanted by *vaccination* when after twenty years of research E. Jenner showed in 1796 that inoculating with matter from cowpox pustulas proved as effective while less hazardous.

Early applications of probability calculus to estimating the effects of smallpox on mortality were made before Jenner's discovery, at a time when inoculation was indeed a hazard. This has to be kept in mind.

The first empirical investigation of the comparative death rates from 'natural' and 'artificial' infection was actually conducted by Mather and Boylston in Boston (begun 1721). Daniel Bernoulli [1760] was not aware of its having been attempted when he tried to evaluate, in terms of *mean duration of life*, how advantageous inoculation was. D'Alembert [1761] harshly criticized his approach, insisting on the problem's complexity. Trembley [1796] confronted D. Bernoulli's results with the few empirical data then available, namely two tables of smallpox mortality established in the Hague (given by Lambert) and Berlin (published by Möhsen). Duvillard's study [1806] was generally considered decisive: not only did he have far more observations at his disposal, as well as more sensitive mathematical tools, but by the time he wrote vaccination had replaced inoculation, undermining the argument from the vaccinal risk:

As to the objections raised against inoculation, they were grounded in both the deadly hazard for the inoculated himself, and the danger of causing natural smallpox to spread and contaminate more people; I do not have to discuss them, since, according to medical reports, vaccine does not kill anyone and tends to sweep the infection away; which perfectly reconciles the states' and the citizens' interests.

Analyses et tableaux de l'influence de la petite vérole sur la mortalité à chaque âge, et de celle qu'un préservatif tel que la vaccine peut avoir sur la population et la longévité, par E. E. Duvillard (du Léman), 1806, p. 7.

Laplace [1812] admitted Duvillard's hypotheses and results, insisting however on the dangerous consequences of eradicating one plague, thus causing the population to increase and be exposed to a relative shortage of sustenance. The latter remark suggests that Laplace may have read Malthus [1798]. Far into the 19th century recurrent doubts were to be raised about the reliability of calculations concerning the benefits of vaccination. Villermé argued that vaccination was hardly useful as long as health standards were so low that people would die of all other diseases: "the vaccine's benefits have been over-exaggerated" [1833, p. 56]. William Farr himself shared this view in part: "it is by no means proved that the general mortality under unfavourable sanitary conditions is much reduced by rendering a child insusceptible of one type while he remains exposed to all other types of zymotic diseases" [30th annual report of the Registrar General of England and Wales, 1867, in: 1885, p. 321]. Farr nevertheless undertook, on the basis of English life tables, to find out how much the mean duration of life would be augmented, were phtisia, cancer, and all the miasmatic diseases together suppressed [suppl. 35th report, 1872, in: 1885, p. 311].

Those works are now classics. A look into their methodology throws light on an aspect of etiological research through probabilistic calculations.

1.1. Daniel Bernoulli (John's son, James' nephew) reports having talked about his project with Maupertuis, who thereafter prompted him to write his memoir. The idea was to construct the comparative life tables of the human race in the natural state (in which mankind is ravaged by smallpox) and in the 'nonvariolic' state (assuming mankind were set free from smallpox) [1760, p. 2. Also see tables in appendix].

One shortcoming arose from the lack of empirical data. D. Bernoulli was well aware of the fact, and he called for physicians to keep files on their variolic patients [1706, p. 9], and for parish pastors to note the ages of the deceased in their registers [1760, p. 22]. Short of exact data, he could only rely on conjectures, and swore that he would be his own "most severe critic" [1760, p. 4] the day accurate mortality tables became available.

D. Bernoulli very clearly exposed the principles of his analysis, as Duvillard acknowledged: "this writing, the best indeed published in favor of inoculation, is remarkable for the sharpness of ideas and soundness of the views. . . . " [1806, p. 5]. Unfortunately D. Bernoulli's mode of calculation could not be applied under different hypotheses, as noted by Trembley [1796, p. 21]: it does not allow for the variation of smallpox mortality and morbidity according to age.

On the other hand, D. Bernoulli clearly stated what is at stake and emphasized that it is important for states and individuals to measure the consequences of possible lines of actions, so they be apt to make *rational decisions*:

> Should he know exactly all the mean proportions taken from a great number of observations, yet well reckoned and reflected, he could give a complete theory of the chances of smallpox: such a theory would yield the rules any reasonable man must follow.
>
> 'Essai d'une nouvelle analyse de la mortalité causée par la petite vérole, et des avantages de l'inoculation pour la prévenir,' par Daniel Bernoulli, 1760, 2, p. 8–9.

1.2. D. Bernoulli started from Halley's life table (see appendix), known to him from Süssmilch [1760]. The original table did not indicate the number of the newborn, out of which 1000 have survived after the first year. Süssmilch (correctly) assumed that Halley had started out with a newborn population of 1238, the mean number of births in Breslaw for the period under consideration.

Having seen Buffon's tables and observed the high infant mortality they give, D. Bernoulli believed Süssmilch's hypothesis inadmissible, and he made the arbitrary assumption of a newborn population of 1300. This yielded a life expectancy (26 years 7 months) somewhat shorter than the life expectancy deduced from the original table (over 27 years). Apart from that, and the fact that the computation did not go farther than 24 years (age at which smallpox mortality becomes almost negligible), D. Bernoulli's basic figures are those of Halley.

The problem was to find an equation which would permit one to transform the natural life table into the hypothetical table which would obtain, should smallpox be eradicated [1790, §11]. Laplace's formulation of the problem is similar.

The only constraint based on observed facts (again, from Süssmilch's data) was that the sum of all deaths from smallpox should be about 1/13 of the total number of deaths, that is, 100 out of D. Bernoulli's generation of 1300 (he actually got 101).

> It has been ascertained in a long sequence of observations that smallpox takes away the thirteenth or fourteenth part of each generation.
>
> D. Bernoulli, 1760, p. 7.

On the other hand, two crucial facts were unknown as yet:

> I could see that, in order for the idea to be worked out two elementary data were wanted: what the annual risk is of catching the disease, in case one has not had it yet; and what the chance of dying is for those who develop it.
>
> D. Bernoulli, 1760, p. 2.

D. Bernoulli took the probability of an individual of undetermined age x catching smallpox to be $1/n$, and that of his dying, given that he has caught smallpox, to be $1/m$. The joint probability of his catching the disease and dying from it, is therefore $1/nm$.

Then D. Bernoulli reasoned as follows. At age x, there are ξ survivors out of the 1300 newborn of the original generation. Out of those ξ children, some have already had the disease in the course of their early years, and it is practically certain that they will not catch it again; s are left who have not had it, out of which during time dx a number will catch it: $-ds = 1/n\, s\, dx^1$. Of those some will die: $1/mn\, s\, dx$. If $-d\xi$ is the total number of children due to die within the year, the number of those who die from other causes than smallpox is $-d\xi\, s/\xi - 1/mn\, s\, dx$.

This leads to an equation [1760, § 5]:

$$-ds = \frac{1}{n}s\,dx - \frac{s}{\xi}dx - \frac{1}{mn}\,\frac{s^2dx}{\xi}$$

which after integration gives, as a function of the other parameters, the number s of children who, at the end of the period, will not yet have contracted the disease:

$$s = \frac{m}{(m-1)e^{x/n}+1}\xi$$

In order to compute s from ξ and x, one needs to know n and m. Their values are assumed to be constant.

1.3. Two general assumptions or "principles" are stated in the introduction to D. Bernoulli's memoir. First, "*as long as one hasn't had smallpox, one continuously runs the same annual risk of catching it*". Second, "*the chance of dying from smallpox, once attacked, is the same at any age.*"

This introduction, written at the time of publication in 1765, gives a few arguments to show that the basic assumptions are at least *plausible.* D. Bernoulli was aware that the lethality of smallpox may vary with time and place, a fact well known to the medical profession already acquainted with more or less 'malignant' epidemics. Scattered observations had seemed to indicate that the disease might kill roughly one out of 7 or 8 attacked [1760, p. 7], but from one epidemic to another the proportion could go from one out of 3, to one out of 40. Furthermore, an opinion current at the time was that the disease is rare but more deadly among adults. D. Bernoulli (rightly) guessed that many adults may have had smallpox without knowing it, hence were protected without knowing it, and that death from smallpox may only be more shocking in adults than in children.

Finally, he assumed that $1/n = 1/m = 1/8$, a proportion which he thought to be attested in large cities, and which might have to be adjusted to local conditions in smaller cities–such as Basel, D. Bernoulli's "little home land", where doctors told him they would estimate at 1/20 the probability of dying when contracting the illness [1760, §10].

So, at any age, one who has not yet had smallpox would run a 1/8 chance of catching it within the year, and a 1/64 chance of catching it and dying.

The appendix gives the table computed on the basis of such hypotheses, by the above formula.

A comparative table for the natural ('variolic') and the 'non-variolic' states is deduced from the latter, assuming that all those people who would have died of smallpox in the first state, would survive in the second. D. Bernoulli disregarded the fact that survivors would be exposed to other causes of death, a defect to be corrected later by Duvillard.

From the tables it follows that in the 'natural' state, from a generation of 1300, 500 die without having had smallpox, 700 develop smallpox but die from other causes, 100 die of it (half of whom are under 5).

According to Lacroix [1833, §113, p. 201] the idea of dealing with duration of life in terms of mathematical expectation originated with Nicholas Bernoulli (Daniel's cousin). Daniel Bernoulli did not use the words 'life expectancy' in his memoir, but he reckoned that in the 'non-variolic' state the '*mean lifetime*' at birth would be 29 years 9 months (against 26 years 7 months in the 'variolic' state), a gain of over three years. But this brought him to the 'decision-problem': *should* inoculation be undertaken?

1.4. Although D. Bernoulli did not conceal his belief that inoculation is "very useful", he claimed that his intention was merely to "throw some light" onto the question, so it can be considered with the fullest possible clarity.

He realized that the State and the individual may have different views.

Suppose all the newborn would be definitely protected from smallpox: the "*relative gain*" in the number of lives would come close to 1/8th of the 'generation' as it approaches 17 years of age (it's "civil birth" [1760, §12d]), and the same generation would not be cut by a half until the age of 24 years 3 months (against 11 years 5 months in the natural state), i.e., its *probable lifetime* would be more than doubled.

The objection D. Bernoulli makes to himself at this point, is that he got such a result while disregarding the risk involved in inoculation *per se.* He then sets that, in order to protect a generation, mankind would have to pay an annual 'tribute' (why annual, and not once and for all when they are inoculated? D. Bernoulli's calculations are not easy to follow here; the general idea is however clear). The conclusion is that, as long as the 'tribute' remains (even slightly) inferior to smallpox mortality (more exactly, as long as inoculation kills less than 1 out of 9.4), "it is more beneficial than detrimental" for the State to pay this tribute, and the more so the younger the children are inoculated, for then "the loss falls upon socially useless infants, . . . letting the gain accrue to the most precious age" [1760, § 14]. This anticipates Quételet's inquiries on estimating the cost for society of raising children who die before they reach adulthood.

To marginal objections, pertaining to a hypothetical lesser susceptibility of infants to the disease, or a danger of inoculation's contributing to artificially spreading the disease around, D. Bernoulli's answer simply is that "physicians should be consulted" . . . : "Let us learn the nature of smallpox from phenomena, rather than deriving pathologies from hypotheses" [1760, §14, 15].

The unequivocal conclusion is: even though inoculation be costly in taking a number of human lives, it is to the *public interest* to systematically apply it, and the earlier in age the better, at any rate under 5 years of age. "It shall always be geometrically true that it is to the State's advantage to favor and protect inoculation" [1760, §14].

The grounds for such a stand are clearly stated: "as long as one accepts *the principle of the greatest utility for mankind . . .*" ("*le principe de la plus grande utilité de toute l'humanité*") [1760, §14]. The better choice is the one that maximizes the *mean lifetime*, because society wants as many adult and productive citizens as possible.

1.5. The individual's viewpoint may differ notably from the State's. Suppose a person has reached the age of 20 without catching smallpox. Society finds virtually no advantage (said D. Bernoulli) in having him inoculated at an age when the disease has already taken 95% of its total share (this may, of course, be contested in several respects, but this is the author's argument). The person "having reached years of discretion" should then think and choose for himself that action which maximizes his chances, balancing a 1/473 (hypothetical) chance of dying immediately from inoculation and 472/473 of lifelong immunity, against an annual 1/64 chance of dying from smallpox with the necessity of taking that chance each year until he contracts the disease (or dies of something else?):

Any man who has not had smallpox shall meet, each year of his life, the crushing necessity of gambling against 63 others, which one shall die of the sickness, and against 7 others, who shall catch it; he drags on such a sad destiny until he catches the disease. Assuming inoculation takes away 1 out of 473, isn't it worth playing against 472 rather than 63, and having to draw once, rather than being subjected to drawing again each year? Can a wise man hesitate about what he should choose? Yet the alternatives are just those of waiting for natural smallpox or getting inoculated.
D. Bernoulli, 1760, §17.

The force of the argument depends on the accuracy of numerical figures. However, in suggesting that *for the same year* the chances of death are 1/64 (from smallpox) against 1/473 (from inoculation), D. Bernoulli anticipated and forestalled an objection of d'Alembert: "is the price of all years of life the same?" [1760, §1].

1.6. "The illustrious d'Alembert flung himself onto the ring", said Duvillard. D. Bernoulli's memoir had been sent to the Académie Royale des Sciences in 1760 and was not published until 1765. D'Alembert, whose influential position within the Academy is well known, succeeded in publishing his own memoir in 1761, and that certainly was not left to chance. The memoir criticized a solution to the problem of inoculation given by "a learned geometer" [1761, p. 88–92] whose name was not mentioned.

D'Alembert did not discuss the State's viewpoint; he only took up the individual decision problem, arguing that a young adult would be foolish to voluntarily expose himself to the hazard of inoculation, just because he is told that the benefit would be three more years of *mean lifetime*: his present years in his prime being indeed more precious than a hypothetical prolongation of his old age . . .

In the Introduction written for the 1765 publication, D. Bernoulli declared himself "surprised" that "persons of merit" should let themselves be deceived like "the common people" by the notion of *mean length of life*. A slashing remark, given the current urbanity of those days. The weakness of d'Alembert's development evidently comes from his well-known imperviousness to probability theory.

Trembley was blunt: "he substituted for Bernoulli's elegant analysis a mathematical theory so very mathematical, that neither he nor anyone I know could apply it" [1796, §4, p. 20]. Duvillard with more restraint noted "a few errors in the philosophy of the calculus", disconcerting on the part of "such a great geometer" [1806, p. 6].

D'Alembert's general argument nevertheless draws attention to two things of importance: first, when trying to estimate the chances of survival of an individual, the *mean life expectancy* computed for the entire generation is only one element of information among others; second, when solving the decision problem, one might have to take into account not only the *quantity* but also the *quality* of the surplus life expected.

1.7. Towards the end of the 18th century, inoculation had become one theme of 'political arithmetics'. Trembley attempted to adjust the theory to the few facts then known: "one had discussed the matter without clarifying it, because too few accurate observations were available" [1796, §1, p. 17].

Trembley devised a mathematical solution of D. Bernoulli's problem using ordinary calculus and not requiring the ratios $1/n$ and $1/m$ to be constant. Then, using two sets of empirical data giving the age at death and number of smallpox casualties, through *trial and error* he sought what the values of n and m should be for each age, in order for the equations to give the result closer to the observed number (see table in the appendix).

The conclusion of that modest yet efficacious piece of work is that D. Bernoulli was in the right when he assumed that the danger of being infected by smallpox (provided one has never had it) was approximately the same for all ages (although he probably had underestimated it, for the data would tend to suggest $1/n$ to be closer to 1/6 than to 1/8). The value of m however "varies considerably" with age, the lethality of smallpox being a maximum during the first five years of life ($\simeq 1/6$), then rapidly decreasing up to age 11 (1/120), then going up again around the end of the second decade (1/60 at age 20), then finally diminishing (1/133 at age 25)... [1796, §16].

Trembley was well aware of the fact that his result merely indicated that facts about smallpox cannot be deduced from *a priori* principles. He also saw that he did not know whether "the danger of inoculation follows the same law as the danger of natural smallpox"

[1796, §18, p. 37], so that he could not answer the practical question of how old children should be when they are inoculated.

His methodological comments are interesting. Some facts, he says, are observed (death from smallpox). Some are directly *deduced* from observations (probabilities are inferred from frequencies: the "true" value of $1/n$ and $1/m$). Some are more *concealed* (not directly observable) and have to be revealed through mathematical analysis (the exact *influence* of smallpox on human mortality). Mathematical analysis does not make up "*new effects*". It does not depart from facts, it makes clear what actually is there by delineating "the way nature works and operates" [1796, §2].

1.8. Duvillard also thought the task of mathematical analysis, which he took to be an art, is to deduce from observed mixed effects of several causes, what effect each cause would have, were the other causes removed.

> It is often possible, through reasoning and simple calculations, to *deduce from effects* produced by two, three or more joint causes, the specific effect that would obtain, should each cause act independent of the others. From a few known data, one thus gets to discern several as yet unknown. Here however, besides the fact that causes constantly vary in intensity, and that only effects produced within a finite time are given, the known is so intermingled with the unknown that one could not succeed in disentangling all those quantities without the assistance of mathematical analysis.
>
> Duvillard, 1806, Introduction, p. 2–3.

'Effects' known in 1806 were somewhat more numerous that at the time of Bernoulli. Duvillard was acquainted with Geneva's district registers, data collected in Berlin and The Hague on death from smallpox and other causes, and a study on the results of inoculation (Dr. Jurin). He sought "what conditions are required" for observed effects to be obtained. Indeed there is not a perfect agreement between the various sets of data: "there is no doubt that the absolute mortality *caused by* smallpox or other illnesses varies somewhat according to place, time, sex, circumstances: amidst such variations one nevertheless discerns *rules rather uniform and general*" [1806, p. 9]. Probability theory allows the 'deduction' of those causal hypotheses which fit observed effects best. For example, it ascertains a few facts inferred by Trembley, which were contrary to common belief: smallpox is "less dangerous" for adults than for infants, two thirds of a 'generation' will sooner or later get it, when the generation reaches its 30th year only four out of 100 are yet unattacked, etc...

It should be remembered at this point that Duvillard's work came after Jenner's discovery (cf. 1.0). His predecessors had studied the effects of *inoculation*. He concentrates on the effects of *vaccination*.

Duvillard had himself constructed a life table (see appendix). Using a method of calculation partly suggested by Lambert, and taking into account the fact that a 'generation' protected from smallpox would be exposed to all other causes of mortality (an improvement on D. Bernoulli), he extrapolated from his natural life table the hypothetical life table that would obtain in the 'non-variolic' state. Those tables display the "*analysis of effects*" from which he "*deduces*" the "*respective influences*" [1806, p. 8] of smallpox and vaccination on mortality, for each age. Those influences are measured in terms of *mean life expectancy*. Since vaccination is assumed to involve no risk of death, the *influence* of vaccination is of course the influence of *withdrawing smallpox*.

Remarkably enough, Duvillard declared: "*my aim is to expose the implications of facts, and neither to reconcile them with opinions, nor to explain them*" [1806, p. 11]. The 'implications of facts' are stated in the memoir [1806, p. 142 sqq.]: in the 'natural' state, out of a generation of one million children, over 85,000 will sooner or later die of smallpox, and their mean life expectancy is 3.9 years. If all the newborn were vaccinated, the mean life expectancy of the whole generation would be increased by 3.5 years (resp. 28 3/4 and 32 1/4). In the latter hypothesis the total population (assumed stationary) would be increased by three and a half million people.

What Duvillard meant by saying that he is *not explaining* facts, is that his "exact mathematical calculations" would have to be made again if other data came to be known, and nothing guarantees that the results would be exactly the same, so the *general law* he disentangled from facts only gives an '*aperçu*' [1806, p. 11] – an analytical glimpse – of reality, which cannot be universalized. The law of smallpox mortality *does not explain* anything; it unravels and *exposes* the bonds between effects (human mortality) and some of the causes which have a definite *influence* on them (smallpox and other illnesses, vaccination).

Note, however, that the 'Form' of a phenomenon may in a sense be considered its cause, and that knowing 'that things are like that' constitutes a sort of explanation of things. This will be found at the heart of Laplace's view.

1.9. Duvillard's solution to the problem of inoculation was held to be decisive by his contemporaries. Laplace had commented on D. Bernoulli's result in the lecture on probabilities he delivered to the students of the Ecole Normale in 1795 [10th lesson, *Oeuvres*, t. XIV, p. 169]. This lecture became the 'Philosophical essay on probabilities', first published in 1819, then published in 1820 as an Introduction to the *Théorie analytique des probabilités.*

The 1812 edition of the 'Théorie analytique..' already includes Duvillard's result [1812, II, ch VIII, §39, p. 413]. The Essay gives a summary of the smallpox controversy, suggesting that Jenner's discovery of vaccination was decisive and had settled the case. Then it briefly summarizes D. Bernoulli's work and d'Alembert's objections, and proceeds to outline the mathematical solution:

> The simplest way of calculating the advantage that one illness's extinction would result in, is to ascertain from observation the number of individuals of a given age it kills each year, and substract it from the number of the dead at the same age. The ratio of the difference to the total number of individuals of the given age would be the probability of dying at that age within the year, if the illness did not exist. By summing up all such probabilities from birth on to any age, and substracting the sum from unity, the remainder will be the probability of living up to that age, corresponding to the illness's extinction. The series of such probabilities will constitute the Mortality Table, relative to the hypothesis, and from it the mean duration of life may be inferred by the above. It is thus that Duvillard...
>
> Laplace, 'Essai philosophique sur les probabilités', 1819, in: *Oeuvres*, t. VII, p. CIV.

Lacroix [1833, §122–123] gave a detailed account of the (now standard) method of calculation and reproduced Duvillard's results without discussion.

Quételet however, though well informed about the success of vaccination, echoed some of the doubts expressed by his friend Villermé about the possibility of increasing the force of other causes of death when counteracting one [Quételet, 1835, p. 253–4]. The argument involves some ambiguity. Christoph Bernoulli [1841, I, 4, §8, p. 255] said an opinion was rather widespread, according to which vaccine weakens the body and makes people more susceptible to other diseases. On the other hand, Duvillard and Laplace apparently held the pessimistic view that the use of vaccine would produce an undesirable increase in population. Villermé's response to this was twofold. First, one should not fight vaccination out of fear of a population growth, improbable after all, for there are other causes of death. Second, one should observe that vaccine is likely to reduce sharply the number of the blind and deformed, which is in itself a good thing. Everyone discussing the subject reasoned as though the birth rate would remain unaffected by a decrease in the death rate, a supposition contrary to facts, as statisticians were to find out later.

William Farr knew of D. Bernoulli's and Duvillard's works. He applied a similar method to show how far mortality would be reduced, in case phtisis, cancer and all 'miasmatic' diseases were suppressed [suppl. 35th report, 1872]. However, he was very well aware of the variability of biological 'laws', and noticed that smallpox reached its *maximum* mortality *after* inoculation was introduced, and started decreasing *before* vaccination was discovered [2nd report, 1838–9, in: 1885, p. 319]: so that many other causes could have influenced the phenomenon, particularly changes in general sanitary conditions (nay, the disease's own 'genius'). He consequently insisted on the differences between the 'laws of mortality' in 'healthy districts' and industrial cities: cancer is a much more significant cause of death in 'healthy districts' than in Liverpool, where generations thinned out by scarlet fever and phtisis left few survivors to undergo the conditions of old age [1885, p. 305–314]. On the other hand, he also insisted on the remarkable stability of mankind's 'law of mortality': the chance of death, large in infancy, decreases from birth on, to reach a minimum towards age 10, then starts increasing again . . . Farr even sketched a table of the chief causes of death characteristic of each age [suppl. 35th report, 1872].

The question of an 'absolute law of mortality' is crucial when one tries to understand Laplace's views on the problem.

* * *

2.0. In 1795 Laplace was 46 years old. He had not published any of his large works yet, though he had given a number of papers, including two papers on probabilities [1774, 1778] to the Academy of Sciences, of which he had been a member since 1773. He had remained cautiously aloof from political events which were fatal to his colleagues Condorcet and Lavoisier. When the republic decreed that scholars were wanted after all, Laplace, with Lagrange, was called by the National Convention to teach mathematics at the newly founded Ecole Normale.

The tenth and last lecture to the students of the Ecole Normale elegantly summarized his doctrine on probability theory. The text is more concise, and includes fewer digressions, than its subsequent versions. I shall refer to this early version, occasionally indicating the directions of later developments.

One fact is perplexing. When Laplace came to explain to his students how life tables are constructed [1795 version, p. 167], he apparently gave results known to have been published later by Duvillard [1806]; so I must presume either that he had been informed of them privately, or that he retrospectively corrected his first draft (which, I think is more likely). Supposedly the only French tables known in 1795 would have been Deparcieux's [1746, see appendix] and Buffon's [1777].

In his 'Memoir on the probability of causes from events' [1774] Laplace said that problems depending on the theory of chance may be divided into two classes: "*the uncertainty of human knowledge pertains either to events or to their causes*" [1774, §II, in: *Oeuvres*, t. VIII, p. 29].

– case 1: *one knows the cause and is uncertain about events*, as when going to draw a ball from an urn, knowing the exact ratio of black to white balls in the urn.

l'événement alors est incertain, mais la cause dont dépend la probabilité de son existence, c'est-à-dire le rapport des billets blancs aux noirs, est connue.
Laplace, 1774, §II, in: *Oeuvres*, t. VIII, p. 29.

– case 2: *one knows the event while being ignorant of its cause*, as when having drawn a set of black and white balls from an urn, the composition of the urn's content is unknown.

l'événement est connu et la cause inconnue.
Laplace, 1774, §II, in: *Oeuvres*, t. VIII, p. 29.

The construction of life tables is intended to solve a problem of the second kind: while observing the events of the successive deaths of human beings, one tries to figure out their 'law of mortality', i.e. the ratio of death – and life – balls in the Urn of their destiny.

2.1. Imagine each human being has to draw a ball from the Urn at the beginning of each year. In case the ball is black, he must die within the year; if the ball is white, he survives until next year's lottery.

A priori the ratio of black to white balls in the Urn is unknown. In such a case Laplace recommends following Bayes' rule and assuming all the possibilities between 0 and 1 to be equiprobable:

Lorsqu'on n'a aucune donnée *a priori* sur la possibilité d'un événement, il faut supposer toutes les possibilités, depuis zéro jusqu'à l'unité, également probables.

Laplace, 1778, §17, in: *Oeuvres*, t. IX, p. 419.

Laplace in fact did not mention Bayes in either of his 1774 or 1778 memoirs, but he rendered him his due in the historical sketch at the end of the Essay:

Bayes, in the *Philosophical Transactions* for the year 1763, directly looked for the probability that the possibilities indicated by past experiences be included within given limits, and he got it in a fine and very ingenious manner, which is, however, a little awkward. The subject has to do with the theory of causes and future events, concluded from observed events, a theory whose principles I set forth a few years later . . . [1819, p. CXLVIII].

When acquainting his students of the Ecole Normale with mortality tables, Laplace did not use the expression 'life expectancy', although later in the *Philosophical essay* . . . [1819, p. CI] he explained that the mean duration of life is an *expectation* in the mathematical sense.

When dividing the sum of the total number of years lived through by all individuals taken into account in a mortality table by the number of these individuals, one gets the *mean duration of life*, actually found to be $28\frac{1}{2}$ years [1795, p. 168].

There was no reason *a priori* to assume that the ratio of black to white balls in the Urn was 1 to 25, rather than 1 to 2 or 1 to 50; *a posteriori*, in fact, the best estimate seems to be that it was around 1 to 27 1/2.

The procedure specified by Laplace to calculate what he called "*the facility of the possibility of the event*" is Bayes'. Bayes' theorem is formulated in the 1774 memoir [§II] as a 'principle'. Laplace however gives a proof of it in the 1778 memoir [§XV] and makes it clear that *as a principle*, that cause which gives the event the highest probability of occurrence should *a posteriori* be considered its most probable cause.

Principe: Si un évènement peut être produit par un nombre n de causes différentes, les probabilités de l'existence de ces causes prises de l'évènement sont entre elles comme

les probabilités de l'évènement, prises de ces causes, et la probabilité de l'existence de chacune d'elles est égale a la probabilité de l'événement prise de cette cause, divisée par la somme de toutes les probabilités de l'événement prises de chacune de ces causes.
Laplace, 1774, §II, in: *Oeuvres*, t. VIII, p. 29.

The *causes* here are the various urns from which *a priori* the event might have been drawn. Call E the observed event (effect); C_1, $C_2, \ldots, C_k, \ldots, C_n$ the possible causes; $P(E|C_k)$ the probability of the event's occurring, under the hypothesis that cause C_k is the case (i.e., that one draws from the kth urn). The probability of cause C_k, given the event, is:

$$P(C_k|E) = \frac{P(E|C_k) \cdot P(C_k)}{\sum_{k=1}^{k=n} P(E|C_k) \cdot P(C_k)}.$$

In the case of life tables, the most probable cause of the observed events is that 'possibility' of dying which is more likely to give a mean duration of life of $28\frac{1}{2}$ years. It is the Urn in which the ratio of black to white balls is 1 to $27\frac{1}{2}$. If the number of observations were *infinite*, said Laplace, the conclusion would be that the proportion in the Urn is exactly that. When a mere finite number of data is available, the thing to do is calculate the probability for the observed mean lifetime to depart from the "true duration of the mean lifetime" only within certain limits. An outline for such a calculation is given in the *Théorie analytique*... [1812, II, ch. VIII, §35].

2.2. Let us take for granted, on the basis of observations available in 1795, that each person annually determines his fate by drawing from an 'Urn-of-death' containing an infinite number of black and white balls in the ratio of 1 to $27\frac{1}{2}$ – or equivalently, that every human being has each year a $\frac{1}{28.5}$ chance of dying. The "absolute possibility" of the simple event consisting in drawing one black/white ball is what Laplace called "*the cause*" of observed effects [1778, §XIV, p. 415]. The 'absolute possibility' is *a priori* unknown and *a posteriori* determined through "*that part of probability theory which consists in going from events back to causes*" [1778, §XXI, p. 439] using Bayes' method.

Pour avoir la probabilité de l'existence d'une cause quelconque... résultante d'un événement donné, il faut déterminer la probabilité que cette cause ayant lieu produira cet évènement, et diviser cette probabilité par la somme des probabilités semblables... relatives à toutes les causes qui peuvent le produire.

Laplace, 1778, §XV, in: *Oeuvres*, t. IX, p. 417.

It is tempting to call a *formal cause* the *absolute possibility* from which the statistical *form* of observed events is derived. This could be written $P(D)$: absolute possibility of death.

Duvillard implied something of the sort when observing that he conducted his analysis going from observed mixed effects to their causes (cf. 1.8), and found out that a human being who is destined to catch smallpox determines his fate each year by drawing from an Urn in which the ratio of black to white balls is 1 to 2.9, whereas an individual protected against smallpox determines his by drawing from an Urn in which the proportion is 1 to $31\frac{1}{4}$.

This however raises a problem, insofar as it points to differences in chances of death. And it is somewhat puzzling to hear Laplace talk about the 'true' possibility of dying, as though a large number of observations were going to allow an even more precise approximation of "truth", and could reveal the "*true law of mortality*" in case the number of observations were infinite.

On conçoit que la précision de ces résultats exige que l'on considère un très grand nombre de naissances; mais l'analyse des probabilités nous montre qu'ils approchent sans cesse de la *vérité*, avec laquelle ils finissent par coincider, lorsque le nombre des naissances considérées devient infini.

Laplace, 1795, in: *Oeuvres*, t. XIV, p. 168.

The later version of the same text is even more explicit:

La probabilité augmente à mesure que l'on considère plus de naissances, en sorte que les Tables représenteraient exactement *la vraie loi de la mortalité*, si le nombre des naissances employé devenait infini.

Une Table de mortalité est donc une Table des probabilités de la vie humaine...

Laplace, 1819, in: *Oeuvres*, t. VII, p. CI.

What is the "true" law of mortality? A few years after Laplace's death [1827], William Farr got (for England) a mean *annual risk* of death, of about 1 to 39 [5th Annual Report, 1841, in: 1885, p. 456]. At the end of the 19th century J. Bertillon found (in Paris) from 1 to 49, to 1 to 54 [Bertillon, 1906, p. 8]. The same risk, in the same European countries, in the seventh decade of the 20th century, was something like 1 to 74...

Of course I have oversimplified things, since Laplace knew the *annual risk* to be different for each age. The theoretical problem however is unchanged, whether one considers the mean life expectancy for each age, or globally the mean life expectancy at birth: for the life expectancy of a 5 year old child, according to Duvillard's estimate, was 43 years 5 months (a figure mentioned by Laplace); William Farr found this to be 48 years [ibid.], and it is now set at around 65 years in France [I.N.S.E.E., 1973].

"The heavens themselves, notwithstanding their orderly movements, are not inalterable", said Laplace [1820, p. CXXI]. It is hard to believe that a man who theorized about the genesis of the solar system did not conceive of a possible evolution of the law of human mortality.

2.3. Let us see how Cournot, a few years after Laplace, posed the problem:

If a living being could be shielded from all accidental causes of destruction, whether they bring about a sudden violent death, or they determine illnesses followed by death more or less rapidly, then one could observe the *natural* duration of his life, that is, the duration which is determined by its organization's intrinsic conditions. Such a duration indeed would vary from one individual to another within the same species; but a rather small number of observations would suffice to yield a fairly stable mean, which would be a sort of measure of the species' longevity. It is possible to come more or less close to the conditions just defined, and thus approximately to solve one of the most interesting problems in zoology, namely the problem of comparing various animal species from the point of view of their longevity, and eventually finding the law according to which longevity is a function of types of organizations, connecting of functions, and action of environments. So far this is only very sketchily known.

In actuality, no individuals, or very few indeed, die a natural death brought about by life's fading: they all are incessantly exposed to causes of destruction, that they oppose more or less successfully depending on how strong they are. An atmospheric change which occasions a mere transient discomfort in the young, or perhaps even leave them unaffected, may cause the old to die. Living is in that sense, as Bichat said, a mere resistance to death. Consequently, without even distinguishing between various causes of death, the variations of mortality according to age, sex and other conditions will show a valuable index of vitality's variations. Indeed mortality may decrease, either because the action of destructive causes diminishes in intensity, or because there is an increase in the resistance of vital forces. Indeed too, a burst of energy in vital forces may multiply the dangers at certain ages in life, thus indirectly contributing to an increase in mortality.

As to the specific case of mankind, knowing the chances of mortality is of high importance not only for people working in medicine, administration, economics; it is also of great interest for all of us. It may prevent being upset by exaggerated dreads or expectations in the ordinary course of life; it may facilitate our submission to nature's severe laws.

Cournot, 1843, ch. XIII, §171, p. 310–312.

Cournot's suggestion clearly is that the chance of death might be divided into two chances. His analysis explicitly follows Bichat's [1800] doctrine according to which there are 'natural' and 'accidental' causes of death. One may imagine that human beings actually are submitted to two annual lotteries. An Urn *Ax* would fix the probability for a 'natural' death within the year for age *x*; for all the people having drawn a white ball from Urn *Ax*, a second drawing from Urn *Bx* would then assign the chance of 'accidental' death within the year for age *x*. It might be fancied that the ratios of black to white balls in the Urns of the *A* series would be constant ('absolute' law of mortality, characteristic of the species), and the composition of the Urns of the *B* series would vary with times, places, sanitary policies, etc . . .

Cournot did not make the first assumption. Much in the spirit of Bichat he thought that probabilities of the first kind are one (phenomenal) aspect of the (actual) *life forces* which *resist* destructive environmental *causes* (manifested through probabilities of the second kind). He however allowed that the "*energy of vital forces*" may vary; and the influence of environmental hazards may also vary. Nevertheless, they must not vary greatly, for Cournot ended up saying that their knowledge will "facilitate our submission to nature's severe laws".

Causes of the first kind are what would be left (said Cournot), were all accidental causes of destruction removed. The hypothesis is immaterial, in the sense that it is impossible to remove *experimentally* all causes of destruction. Bichat [1800, I, art 10] claimed physicians hardly ever observed genuine deaths from old age. Studies like Duvillard's however showed that mathematical analysis can surmount the empirical difficulty of separating the two kinds of causes. What is remarkable in this situation, is that the 'absolute' probability of a 'natural' death, rather than being *the* cause of human mortality, became a *residue* that the International Classification of causes of death soon ranked under the heading "ill-defined causes"–the word 'cause' then departs from its Laplacean (first) sense and takes on another meaning.

Laplace himself never suggested he would adhere to Cournot's distinction. Thus, the question how he interpreted what W. Farr nicely termed "*the law which regulates the waste of life*" [5th Annual Report, 1841, in: 1885, p. 455] is still unanswered. Should it be understood that the 'true' life table, of which existing tables give only rough approximations, would be the Table constructed at the end of time by an Intelligence capable of recapitulating all the durations of all lives of individual human beings, thus giving the *mean lifetime* of *homo sapiens*? Should it be understood that the 'absolute' possibility of human life and death is determined by the universe's structure and underlying eternal laws, the observed differences in mortality ratios being the result of successive random samplings from the Urn-of-death?...

2.4. Let us return to the 1795 10th lesson on probabilities.

It is clear that the mean duration of life would be augmented if wars became less frequent, if life standards were higher for more people, and if through whatever means, man succeeded in rendering the soil he lives on more salubrious, and in lessening the number and seriousness of illnesses. That is what was done about smallpox... [1795, p. 168–169].

So, Laplace was perfectly aware of the fact that the 'absolute' possibility of death may be changed when people are systematically inoculated. For his students he warmly praised the work of Daniel Bernoulli, "who found, through an ingenious application of probability calculus, that inoculation would sensibly augment the mean lifetime" [1795, p. 169]. In the same lecture he drew attention to a line of research the implications of which are significant: one should inquire more precisely, he said, into occupational life tables.

The various conditions of life offer, as regards mortality, very noticeable differences, relative to fatigues and dangers inseparable from each condition; it is necessary to take these into account in computations based on the durations of life. These differences however have not been sufficiently established. They will be some day: then it will be known what sacrifice of life each occupation demands, and such a knowledge will be taken advantage of to alleviate their dangers.

Laplace, 1795, in: *Oeuvres*, t. XIV, p. 168.

Those factors which prolong or cut short the mean lifetime are factors *influencing* mortality, that a probabilistic analysis can detect.

A comprehensive statistical study of occupational mortality was made by J. Bertillon at the end of the 19th century. He found (I simplify) that the annual death rate for men aged 50 to 59 is, out of

1000: 40 for painters, 30 for masons, 20 for nursery-gardeners (the latter proportion being close to the proportion for the general population).

One may reason about such facts in the way Laplace himself said one should reason about a difference between death rates observed in different places: first make sure that the difference is not attributable to chance, by computing the probability of a *special cause* being *indicated* by the facts; second, look for the cause of the observed difference.

The soil's greater or lesser salubrity, its elevation, its temperature, the inhabitants' customs and the governments' operations, have a considerable *influence* on mortality. But the investigation of *the cause of observed differences* should always be preceded by the *investigation of the probability with which such a cause is indicated.* Thus the ratio of the total population to the annual number of births, that we know in France is 28 1/3, does not come to 25 in the old duchy of Milan. Both ratios having been ascertained from a large number of births, the existence in the Milan area of a *special cause of mortality* cannot be doubted. It is of importance for that country's government to investigate that cause and have it removed.

Laplace, 1819, in: *Oeuvres*, t. VII, p. CII.

Let us apply the same line of reasoning to occupational mortality tables. Nursery-gardeners are not very likely to determine their fate by annually drawing from an Urn other than the common Urn: observations are compatible with the hypothesis that their mortality is but a random sampling from that Urn. It is however extremely improbable (provided the number of observations is sufficient) that painters would determine their 1 to 24 fate by drawing from an Urn in which the ratio is 1 to 49. This is a consequence of James Bernoulli's theorem. Hence one should, as before, determine the most probable Urn from which the painters are drawing (Bayes's procedure), and the *difference* between the possibility of dying in the general population (or: from all causes) and the painters' possibility of dying *measures* the *influence* of the *special cause* to which the painters' excess mortality is *due.*

The need for some kind of standard of reference against which the specific painters' mortality should be judged is apparent here. Hence the notion of an 'absolute' possibility of death. Call C the special cause of the painters' overmortality, $P(D_{50})$ the possibility for a human being to die in his fifties. The situation can be schematized:

$$P(D_{50}|C) - P(D_{50}) = \epsilon$$

ϵ measuring the difference, i.e. the *influence* of the *special cause.*

2.5. On how to go about identifying the *special cause*, Laplace did not say much. Cournot's suggestion [1843, §102] is that *prima facie* the special cause in question would simply be the fact of being a painter rather than a non-painter, i.e. the characteristic property defining the set of individuals having the observed 1/25 death rate. In other words, the total set of observations would be 'cut' according to the criterion painter/non-painter, and the 'cause' of the observed difference would simply be the given criterion for the corresponding subset. Conceivably medical notions would then lead to re-cut the total set according to the criterion: manipulating lead/not dealing with lead, and so on.

Laplace himself allowed that causes so designated might be of varied natures: sickness (smallpox), occupation (painter), climate (Milan), etc . . . What these have in common is that they *influence* the chance of death with enough *steadiness* for their effect to be ascertained throughout a long series of observations. Due to the 'power' they have of influencing the chance of death, one is inclined to consider them as *efficient causes.*

The question is blurred, however, if one reflects that the 'absolute' possibility with respect to which the difference is ascertained, may itself vary while one is measuring the difference. When W. Farr found that in England in the 1840's, "the mean duration of life, technically called the expectation of life" [5th Annual Report, 1841, in: 1885, p. 456] was 41 years, was he entitled to conclude, along Laplace's lines, that there had been, at the time of Laplace, a special cause responsible for the fact that, what Laplace believed to be an approximation of the 'absolute' possibility of death, deviated widely from the 'true' possibility?

2.6. Laplace's 'true' thought about this is hard to grasp. It does not seem that he wanted to stick to a strict causal principle, for he said that we attribute similar effects to similar causes, *and conversely*, only on analogy:

C'est par l'analogie que nous attribuons des effets semblables à la même cause ou à des causes semblables, et réciproquement; ainsi nous jugeons que des êtres pourvus des mêmes organes, exécutant les mêmes choses et communiquant ensemble, éprouvent les mêmes sensations.

Laplace, 1795, in: *Oeuvres*, t. XIV, p. 151.

At the level of what I called *efficient causes*, Laplace did by no means exclude the possibility that different causes may produce the same effect (the same drop in mortality rates), or even that similar causes (inoculation) may have various effects (may be more or less deadly). The only methodological requirement, when discordant effects are observed which were believed to proceed from the same cause, is to establish the fact of a constant discordance, and to find a *special cause* that provides an account of it. It is unimportant whether the same cause is always involved in the same difference, what is crucial is that the specified cause render a complete and exact account of the observed difference.

Going back to the Urn image: from the same Urn, two series of drawings will be expected to give (at least slightly) different results (same cause, different effects); conversely, the same observed mortality rate may (within certain limits) be drawn from non-indentical Urns (same effect, different causes). One is entitled to raise a question when two sets of mortality rates have been ascertained to be more probably drawn from different Urns than from one and the same. Then the cause should be sought, which explains (– what exactly?) the substitution of one Urn for the other, or the bias of one set of drawings from the common Urn... Understandably, qualitatively different causes may produce the same biases, and (within certain limits) two different biases may result from the same cause, etc...

I think that tells why Laplace chose to state a "*principle of sufficient reason*" [1819, p. VI] without ever stipulating anything like the principle of "*same cause, same effect*".

Nevertheless Laplace professed the faith of a determinist. From the 1795 lecture on probabilities on (though without using yet the striking supposition of an intelligence capable of embracing in one formula the movements of all bodies in the universe [cf. 1812, II, I, §1, p. 177] he firmly declared that natural phenomena "*are but the mathematical result of a small number of invariable laws*" [1795, in: *Oeuvres*, t. XIV, p. 152]–the laws themselves being the surface expression of *forces*, i.e. properties of reality. The 1795 statement was retained unaltered in all subsequent versions of the Essay.

All events, even those which on account of their tininess do not seem to hold to the universe's great laws, are as necessary a consequence of them as are the sun's revolutions. Due to our ignoring the actual connections linking them to the entire system of nature, they have been made to depend on final causes, or on chance, according to whether they occurred or succeeded each other with regularity, or without an apparent order; such imaginary causes however have been successively pushed back with our knowledge's boundaries, and vanish entirely in the eyes of a sound philosophy, which sees in them only an expression of our being in ignorance of genuine causes.

Laplace, 1795, in: *Oeuvres*, t. XIV, p. 146.

The *true causes* of events are thus the universe's great laws, from which events *necessarily follow*, in the sense that they are *mathematically derivable* from them.

In particular, the *true cause* of observed mortality rates is the *law of mortality*, i.e. the 'absolute' possibility of the simple event symbolized by one drawing from the death-urn. The regularity of that law is blurred, at the level of phenomena, by the extreme variability of what are commonly called 'causes of death'. It would be manifested (as a consequence of James Bernoulli's theorem) on a long range of observations, when *irregular* (*variable, particular, perturbative*) causes such as wars, epidemics, climates, sanitary policies of states, would have "mixed their influences" and "neutralized their respective effects". Results taken from a very large number of observations point to "the truth" [cf. 1819, p. LVI].

Laplace declared himself struck by the fact that such a regularity arose even in "effects due to moral causes"; e.g. in the city of Paris the annual mean number of births, or "of letters sent to the dead-letter office, for want of an address" [1795, in: *Oeuvres*, t. XIV, p. 161–2]. *A fortiori* the fact that in the same city similar percentages of deaths in the young, the adult and the old recur each year, should be expected: it merely displays "*the development of the respective possibilities*" [1795, p. 161] of *constant causes.*

In an indefinitely prolonged series of events the action of regular and constant causes must finally prevail over that of irregular causes.

Laplace, 1819, in: *Oeuvres*, t. VII, p. XLVIII.

Note that causes which are considered anomalous from the point of view of death rates may be viewed as *constant* in their own respect. Thus Laplace holds that there is a regularity in climatic variations which 'disturb' mortality. Irregularity lies in the collision of effects rather than in causes.

The problem remains of determining to what extent the postulate of an *absolute invariability of constant causes* can be dispensed with, that is, of knowing whether the distribution of 'possibilities' over simple events is a fixed feature of the Universe, or is subject to (eventually random) variations. In the first case, one may not assume that "past events *have an influence on the possibility* of future events" [1795, in: *Oeuvres*, t. XIV, p. 162]–that is, past events have no more real *influence* on future events than one drawing from the Urn on the following, the future can be predicted,–there *is no history*. In the second case, as may be conceived, the way is open for what I called *efficient* causes to actually alter (and not merely conceal) the 'true' ratio of black to white balls in the Urn, thus inducing irreversible changes and a historical drift.

2.7. When Laplace explained how the "influence of an event on the probability of any future event" should be calculated [1778, §XXIX, and 1812, II, VI, §32], he assumed the (unknown) possibility of the simple event to be stable, and the future event is to be predicted from the observed past event *via* the computation (through a Bayesian procedure) of their common 'absolute' possibility. There is no assumption of an actual influence of what has already occurred, on what is going to occur. There is an influence of what has already occurred on our capacity to evaluate the 'true' possibility, hence to accurately predict what may obtain next. "*For the happening or failing of an event in different trials are so many independent events*", said Thomas Bayes [1763, Section I, prop. 7, p. 384].

In the 1778 memoir Laplace stated what he considered a "quite remarkable" theorem in its simplicity: "the probability of a subsequent event similar to one previously observed is in the ratio of 1 to $\sqrt{2}$ to the probability of the observed event, as determined by using those possibilities of simple events which result from data" [1778, §XXIX, in: *Oeuvres*, t. IX, p. 475]. When generalizing this theorem he remarked that the probability of future events decreases as the prediction ranges farther.

Assume, for instance, that the possibility of infant mortality (under one year) in 1820 has been estimated at .19 (from an observed rate of 90 out of 1000, given by J. Bertillon [1906, p. 117]). The probability for the same rate to be observed during *n* consecutive more years decreases as *n* increases, because the "slight error" that may have been committed in estimating the actual possibility from the observed rate "piles up in proportion of the number of events compounding the future event" [Laplace, 1778, §XXIX].

Laplace's favorite examples were the tossing of a coin [cf. 1795, in: *Oeuvres*, t. XIV, p. 159] or dice and the players' skill [1778, §XIV]. I believe that the first Laplacean version of this [1778] came directly from James Bernoulli, who in his 'letter to a friend on the plays of the *Jeu de Paume*' treated the players' skill as a *cause* [posth. 1813, paginated apart, p. 3].

What Laplace wanted to emphasize was that, in case the coin is fair (or if a different coin is tossed each time) it does not make sense to try to guess future events from past events, not because the outcomes *are* random, but rather because the possibility of the event is known *a priori*, and it would be stupid to set out to predict future events on the basis of a (necessarily incorrect) *a posteriori* evaluation of a possibility which is *a priori* exactly known. "*Past events then have no influence on the probability of future events*" [1778, §XIV].

Quels que soient les évènements déjà arrivés, *leur possibilité absolue reste toujours la même*, ce qui rend la considération du passé entièrement inutile lorsque cette possibilité est exactement connue.
Laplace, 1778, §XIV, in: *Oeuvres*, t. IX, p. 415.

When however the possibility is *a priori* unknown (unfair coin, skilled player), the outcomes are not *less random* in themselves, in the sense that they depend on the actual 'absolute' possibility of the event just as before, but we *learn from experience* and *a posteriori* improve our guess as to the actual possibility. Let *tails* come out more often than *heads*: "that in itself will lead us to believe that there is in the make of the coin *a constant cause in its favor. Past outcomes then do have an influence on the probability of future events*" [1795, in: *Oeuvres*, t. XIV, p. 159]. They do *not*, however, have an influence *on future events*. Probabilities are, according to Laplace, what human knowledge grasps of the actual determination of events [cf. 1819, in: *Oeuvres*, t. VII, p. VII].

Note that, when pointing to a *constant cause* of the observed effect, we explicitly refer to a difference between what *should* obtain under 'regular' circumstances, and what actually obtains. Laplace held that regularities always prevail in the long run, so that in his perspective knowing the causes of irregularities was a side-issue anyhow.

2.8. To conclude in what appears to be a Laplacean spirit: as they summarize more and more observations, life tables supposedly develop the 'true' possibility of death for each age, and that possibility is what we know of (something like) the extent to which *life forces* resist *physical forces* (as Cournot said).

Knowledge progresses, forces persist.

> Let us not uselessly and often dangerously try to resist the inevitable effects of the enlightenment's progress.
>
> Laplace, 1819, in: *Oeuvres*, t. VII, p. LXXVIII.

Suppose that 'enlightenment's progress' should result in modifying forces themselves, and the 'absolute possibilities' derived from them?

Such an hypothesis was suggested by Quételet when he underscored the fact of a dramatic decrease, within a century, of the annual chance of death (from 1/30 to 1/40). In his ordinary manner he recapitulated all the causes *influencing* death rates, first *natural* causes (such as sex, age, season . . .), then *perturbative* causes (such as wealth, enlightenment, 'morale' . . .), and then he clearly stated that "*the more influential causes*" [1835, t. I, ch. VI, p. 207] are of the second kind, i.e. the progress of wealth and civilization, including the decline of the Church (she "*depressed our vital forces*") [cf. 1835, t. I, p. 246].

It is not clear in this formulation what exactly the 'absolute' death rate should be, or whether one wants one. Probability theory needs the postulate that there are recurrent 'simple' events (of which death is a beautiful paradigm example). It does not *more* (nor *less*) require the possibilities of simple events to be fixed for eternity *than does medicine require 'morbid entities' to be stable in all their manifestations in order for physicians to diagnose them.*

Condorcet [1785] tried with more subtlety (Todhunter said, confusion) than Laplace to sort out methodological postulates.

Commenting on the method set up by Bayes and Laplace for "determining the probability of future events from the law exhibited by past events", he took as an example the question of finding what the probability is for one of two 'opposite' events (e.g. *head* and *tail*) to occur next, given the fact that it has occurred, say, 100 out of 150 times in previous trials.

> The question implies that the probability of such events constantly stays the same throughout the trials, i.e. that the unknown law determining their production is constant. In fact, should that condition not be fulfilled, we know from the calculus as well as from mere common sense that the probability for either event to obtain in the future is the same, no matter what their sequence was in the past.
>
> However we also know from the calculus what the probability is of there existing a constant law governing the production of events.
>
> Condorcet, 1785, p. LXXXIII – LXXXIV.

In the following Condorcet outlines the results he had arrived at in Section 4 of his 'Mémoire sur le calcul des probabilités' [1783].

In fact, Laplace's methodology of causal research apparently can be detached from any postulate pertaining to the existence of 'absolute' causes, at least when it is stated in Bayesian terms. For it seems that the Bayes' and Bernoulli-inverse procedures are not equivalent in that perspective, at least as Laplace described them.

> For instance, in order to identify the best of several treatments in use for curing an illness, it is sufficient to test each of them on a same number of patients, and render all circumstances perfectly similar. The superiority of the most advantageous treatment will be manifested more as the number of patients will grow larger, and the corresponding probability of its being advantageous, as well as the extent of its superiority over others, will be shown by calculus.
>
> Laplace, 1819, in: *Oeuvres*, t. VII, p. LXXVII.

Now, should it be the case that, in the course of experiment, the efficacy of one particular treatment dropped sharply, would one want to consider its overall 'mean' efficacy, or look for a 'special cause' of the variation?

In his lecture to the students of the Ecole Normale, Laplace explained that the "influence of past events" was to be determined by the means of the following *principle*:

> If the respective probabilities of the observed event and of an event compounded of the former and an expected event, are calculated *a priori*, the second probability divided by the first will be the probability of the expected event, given the observed event.
>
> Laplace, 1795, in: *Oeuvres*, t. XIV, p. 158.

In short:

$$P(E \mid O) = \frac{P(O \cap E)}{P(O)}$$

This is nothing else than a definition of conditional probabilities, from which Bayes' theorem is derived. It is compatible with a 'subjective' *a priori* evaluation of the events' probabilities, as well as with further adjustments of the estimate, as new effects develop.

When however Laplace came to explain how to look for "the clues that may guide our conjectures about causes", i.e. about possibilities governing the development of events, he stated James Bernoulli's 'inverse' theorem (without naming its author):

> If observations or experiments are indefinitely multiplied, their mean result converges towards a *fixed* term, so that, as small an interval as wanted being taken around that term, the probability for the mean result to fall within the interval will finally differ from certainty only by a quantity less than any assignable magnitude. That term is *the truth itself* in case positive and negative errors are of equal facility.
>
> Laplace, 1795, in: *Oeuvres*, t. XIV, p. 161.

The assumption that death rates converge towards a *fixed term* (the *true* possibility of death) does not seem necessary to an analysis of either 'constant' or 'variable' actual causes of death.

2.9. The difference between 'constant' and 'variable' causes is itself unclear, as shown by frequent slips in Laplace's vocabulary. For the "variable and unknown causes we comprise under the word *chance*" are the very causes which through their effects' irregularities finally shape the event's 'absolute' possibility, while the cause of a variation may be looked for only if it is manifested with a certain *constancy* and the variation cannot be "attributed to chance" [cf. 1819, in: *Oeuvres*, t. VII, p. XLVII sqq.]. The important distinction is thus not between constant and other causes, but between the probability of an event (e.g. death) *from all causes*, and the probability of the same event from a specified cause (such as an illness).

The questions how to specify pertinent causes, and how to evaluate chances, may in some respects be considered technicalities, as suggested by Cournot.

In a chapter on 'the variability of chances', Cournot [1843] said that in the (frequently encountered) cases in which "the chances of an event are of such a nature that they change from one trial to another, or from one series of trials to another", the event's probability is reducible to a probability compounded of the probability of drawing a white ball from a certain Urn, and the probability that the given Urn be selected among a large number of Urns. One of the tasks of analysis is to find out whether the Urn was more probably selected at random, or by "non-fortuitous causes". In the hypothesis of random selection, the probability of extracting a white ball from the selected Urn converges towards the mean value of the probabilities of extracting a white ball from either Urn in the lot. In the hypothesis of non-random selection, one may assume that the Urn from which each drawing is to be made, is itself specified each time by a previous selection, and so on. As the theoretical setting gets more complex, the number of trials to be conducted, in order to arrive at a conclusion, increases; but it can be computed. However, "in case the conditions of the selection or drawing should be submitted to progressive variations in time, one might never reach a stable mean value" [cf. 1843, ch. VII, p. 146].

This is exactly what is observed in the case of life tables. Cournot ascribes the phenomenon to an underlying variation in the "vital forces" or "the energy of causes of mortality". At some ultimate level, one always ends up saying: 'here is how things are'.

> If a large number (e.g. 10,000) of newborn children were selected at random, and followed up to the instant of their death, and the number of survivors at each age were entered with each age, one would get a mortality table, from which a table of the probabilities of human life could easily be inferred. Such a table would carry the kind of fortuitous errors affecting any probability table deduced from a finite number of observations, even when the chance setting does not vary throughout the experiment. It would be further affected by sudden and irregular variations which arise within causes of mortality in the course of experiment. For instance an epidemic which broke out during the twentieth year of the experiment would result in a perturbation for age 19–20 in the table. Furthermore, if within the lapse of time required for the table's completion, causes of mortality had undergone slow progressive variations, all the numbers in the table would be affected. And the portion of the table embracing advanced ages would not correspond to the portion embracing infant and young years.
> Cournot, 1843, ch. XIII, §172, p. 313.

Statistical offices in Cournot's time were already familiar with such facts. William Farr, for example, conceded that probabilities of life, as inferred from mortality tables, far from being 'absolute', are always relative to a set of circumstances: "the life table is an instrument of investigation; it may be called a *biometer*, for it gives the exact measure of the duration of life under given circumstances" [Construction of life tables . . . , in: *Transactions of the Royal Society*, 1859, p. 838–841].

The task of defining relevant subsets in such a way that significant differences in mortality rates point to *special causes*, lies outside the scope of probability theory. Cournot [1843, Ch. VIII, § 102] contemplated the possibility for an experimenter-statistician systematically to try all possible 'cuts' on the set of data. He commented that in such a case probability theory allows the prediction that there will be 'anomalous' results due to chance rather than to special causes. His conclusion was that it is quite natural to cut the set of statistical data according to preconceived categories, provided one has criteria for discarding certain hypotheses after they have been tested.

In the case of mortality data, one of those systems of categories was to be the international classification of causes of death. This is a list of what Laplace would have called *special causes* (what I have called *efficient* causes), that is, diseases or casualties significantly curtailing the *mean life expectancy* of the people they afflict. The fact that *old age* is not one of the causes on the list should not surprise us at this point, insofar as dying of old age ultimately implies having escaped all accidental causes of destruction, and in Cournot's and Bichat's sense, having carried out the 'natural' probability of survival (*formal* cause) granted to the human species.

If one does not want to assume that this is an 'absolute' possibility of death determined within the universe's structure, one may want to consider that there simply exist individual living beings who are exposed to external (or internal) hazards and who each individually die their own death.

* * *

3.0.

Philosophical spirit, that is, reason applied to society's interests and concerns, has been slandered so often, that one must be permitted to recall obvious facts speaking in its favor. One of the most striking is the bettering of the conditions offered to the ill in hospitals, due to efforts of men who were known to be moved by mere human philanthropy. Through their pains, even though in a difficult time, the practice of cramming several patients in one bed at the Hotel Dieu was abandoned, means were taken towards cutting down the terrifying mortality of foundlings, district hospices were created for child-birth and nursing, and many more institutions working towards public order and well-being, that it would take too long to enumerate here. During our highly praised 17th century, the mortality at the Hotel Dieu in Paris roused the indignation of foreigners: 3000 patients a year perished for want of care (See the *Essay tending to prove that in the Hospital called* l'Hostel-Dieu *at* Paris, *there die above* 3000 per annum, by *reason of ill accommodation* (1687), several Essays, etc., by William Petty, edit. of 1755, p. 63).

Religious zeal and pity may indeed provide assistance; only reason, however, can dispense it with equity and institute an orderly economy, so that the greatest number of people may have their share, and that the future may be rich in the resources prepared for in the present.

Lacroix, 1833, §124, p. 223–4, note.

Medical causes of death are but one kind among the many (poverty, lack of sanitary policies, environmental nuisances) taken into account by statisticians. From a statistical point of view, being a male, a coal-miner, a resident of an industrial district, are factors which cut short life expectancy just as developing phtisis does. The rational procedure is to first *assess*, then *measure*, their causal influence.

What is peculiar about *medical* causes of death is that they are (now) diagnosed by the medical profession before being compiled for statistical purposes. This phenomenon is recent in history. The 1837 English Registration Act, for example, required the informant (usually a member of the family) to declare to the local Registrar both the fact of death and the accident or disease causing it: the obligation to produce a medical certificate was imposed much later.

There obviously can be no mortality statistics without 'positive' diagnoses of deaths (whoever makes them) and no cause of death statistics without someone's making 'etiological' diagnoses.

Physicians who initiated the cause of death statistics in the course of the 19th century were aware of participating in a collective effort of human reason and civilization to scientifically detect, classify, evaluate and finally control the plagues by which the human race was thinned out. They had realized that the *waste of life* characteristic of human mortality is partly *unnecessary*. "Death is inevitable, said W. Farr, but why is life cut short? Why do the people of England live only a mean life of 41 years?" [30th Annual Report, 1867, in: 1885, p. 258]. These physicians insisted on the necessity for death and its causes to be certified by qualified persons, in order that such questions might be answered rationally.

If Laplace was right in saying that probability theory only makes precise what people with a sound judgment feel instinctively [cf. 1819, in: *Oeuvres*, t. VII, p. CLIII], one may hypothesize that diagnostic procedures involve a rationality not entirely foreign to the rationality of statistical procedures.

3.1. Telling a dead body from a living one is not a very difficult task in general, and until the 1800's most human societies did not resort to specialists for that. The institutionalization of certification by medical practitioners was however officially justified by the concern that nobody should be buried alive. Medical students were then supposed to learn how to discern *apparent* from *real* death. A number of books on the 'signs of death' were written during the 19th century. One of the most important of them is by E. Bouchut the *Traité des signes de la mort et des moyens de prévenir les inhumations prématurées* [1849]. Another monumental work is P. Brouardel's [1895]. I shall not insist on the partially outdated contents of their teachings, for it is well known that the criteria of death were redefined in the early 1970's [Hastings Center]. Methodological principles, however, have not changed.

Bouchut distinguished between two sets of signs. 'Remote' signs (of the decomposition of the corpse) are obvious even for the uninformed. 'Immediate' signs (of the cessation of vital functions) are not easily differentiated from signs compatible with apparent death.

Brouardel insisted that no one sign is pathognomonic, that death is pronounced on the basis of a set of converging indices. In a lecture to his students [1895, I, ch. II] he mentioned several cases of people considered dead following asphyxia, lightning, blow on the head, freezing, intoxication, etc . . . , who revived "on the mortuary's slab". In 1784 the French Academy of Medicine had declared that "signs of death are certain enough, so that mistakes are *impossible*, provided death is always verified by a physician, and the physician has a sufficient ability to diagnose it" [ibid.]. Brouardel was not so positive: "even though *the probabilities are very high*, that nobody is buried alive, given the circumstances in which deaths are (or rather, are not) verified, it is nevertheless impossible to claim that such a dire contingency cannot occur" [ibid.]. He particularly emphasized the fact that the physician must be "capable of interpreting the value of signs, and *make a genuine diagnosis, just as he would do for an illness*" [ibid.]. One sign is not sufficient, a negative sign even less so, and "during the first hours, there are but negative signs": the body is insensitive, motionless, heartbeats are not perceived . . . One might say: just *wait* for the positive signs of decay to appear, but *waiting* may be precisely what may cause apparent death to turn into real death. Brouardel therefore minutely reviewed the various signs to be checked, and concluded:

The signs of death taken jointly give us a quasi-absolute certainty that the body is dead–that is, if we are physicians, if we have thought about the question, if we have carried out a diagnostic procedure that is neither commonly performed, nor taught in hospitals. Yet I hold it good to stick to a philosophical doubt.

P. Brouardel, 1895, I, ch. III, p. 54.

The *positive diagnosis* of death is a probabilistic judgment.

3.2. A course like Brouardel's was intended to provide medical students with scientific bases (relative to current scientific knowledge) for their diagnoses. The course catalogued all possible signs and discussed the 'value' of each, particularly insisting on cases in which a sign is observed in death's absence (e.g. hypothermia of alcoholic intoxication) and cases in which a sign is missing in death's presence (e.g. rosy complexion in CO asphyxia).

The positive contribution of such medical school teaching is best described as giving an estimate of the probability for certain signs to appear, when death is actually there, and for those same signs to appear when it is not. The numbers are not worked out, but there are 'good', 'better' and 'poor' signs, hence an implicit quantification of probabilities.

When, having memorized the signs, the physician is called to certify a presumed case of death, he must first identify the signs in the particular case, then judge of the probability (given the signs observed) for death to be 'real and constant'.

Identifying signs is hardly a trivial task, even though death is a less elusive event than are many illnesses (what is said here would *a fortiori* apply to the diagnosis of any morbid state). It may even be a somewhat disconcerting experience for a practitioner to come across a symptom that looks 'just as it is described in the books'. The point to be made here is, that a correct identification of signs is not enough to constitute diagnosis, although there would not be a diagnosis without it.

In order to judge of the (conditional) probability for death to be real, the certifier takes into account the probability for each type of sign (or the set of signs) to be manifested in the presence or absence of death in general:

$$P(D \cap s), \quad P(\neg D \cap s)$$

He recasts these probabilities for the particular case of this individual death D_i on the basis of any available information (case history) bearing on the value of presently observed signs s_i:

$$P(D_i \cap s_i), P(\neg D_i \cap s_i)$$

and weighing the probability for the presently observed signs to accompany death against the probability for the same signs to be present on any alternative, he (so to speak) evaluates the probability for the individual to be actually dead, given the observed signs:

$$P(D_i|s_i) = \frac{P(D_i \cap s_i)}{P(D_i \cap s_i) + P(\neg D_i \cap s_i)}$$

which describes a Bayesian procedure.

3.3. The identification of death on the basis of its signs does not only mean that death is *recognized* through its signs, but also that the pattern of observed signs (*formal cause*) would *at the limit* be *equivalent* to underlying actual death (*material cause*).

Clinical medicine traditionally understands signs as the external expression of an internal structure or state. The 'level' of what is conceived to be the 'internal reality' may vary with medical knowledge, which creates the possibility of a corresponding reevaluation of the signs' significance. A century ago the absence of heart pulsations was a good sign of death, insofar as the stopping of the heart defined the actuality of death. Nowadays death is conceived to take place at the level of cells rather than of organs, and the most indicative signs of deaths are signs of prolonged cerebral anoxia. In either case there is corporeal damage which is diagnosed from the form assumed by clinical symptoms.

The *definition* of death goes along such a line of its formal/material cause, is independent of time (signs are contemporaneous with underlying death), and makes it clear that ideally death *implies* its signs, and observing its signs is a *sufficient* reason to assert death. The *positive diagnosis* of a singular death goes along the same line; the probability (*at the limit*, the certainty) that death is *materially* there is estimated from the pattern of observed signs.

Death (as a *material cause*) is said to *explain* observed signs insofar as they also are its *effects* ('*he is cold, because he is dead*'). The material cause is thus turned into an efficient cause, the drop in body temperature is taken to *follow* brain death, the attempt at *explaining* rather than *defining* carries with it a switch towards conceiving death along the line of efficient causes and succession of events in time. Each step of the underlying process of death is then seen as resulting in a particular outlet of signs. The more detailed the correspondence, the better the explanation. There is no limit assignable to such an explicitation.

In the statistical sense, the explanation goes, so to speak, the other way around. The 'absolute' probability of death for a certain age (formal cause) explains (for the part pertaining to the fact that we are all mortal and that there is a 'law of mortality' of our species) the (material) fact of individual death. In fact, it explains it very little, for the mean probability of dying shortly is very low at any age.

The factual basis of mortality statistics is a set of medical diagnoses of death. It is therefore not improper to say that the 'possibility' of death for each age, which from a statistical point of view characterizes the whole population, is grounded in a multiplicity of individual subjective probabilistic judgments. This is not to say that it has no 'objective' character, but that its objectivity is grounded in the consensus of a scientific community. Again, such remarks would apply even more perspicuously to the case of morbidity statistics.

We shall now turn to etiological diagnoses. A separate study of *positive* and *etiological* diagnoses is evidently a convenience, while in actuality both are intermixed. The lack of a 'good' cause of death may throw doubts on the reality of the event, information on presumed causes is always an important part of the epistemic context within which death is certified.

3.4. After certifying that death has actually occurred, the physician must, according to internationally codified rules, declare what caused it. Example:

death was caused by acute renal failure, probably due to chronic pyelonephritis aggravated by dehydration, and associated with diabetes, atherosclerosis and hypertension.

H. Mondor hospital, 1975, file number 1.02.14.

This of an overweight 73 year-old woman, whose diabetes and hypertension had been known and treated for several years. On April 28, 1975, she was hospitalized because of an episode of complete anuria. On admission to the hospital she was perfectly conscious and declared that she just felt tired and weary. She actually had lost 20 pounds in the previous few months, and had recently had abdominal pain, with nausea, vomiting and profuse diarrhea.

The *first etiological hypothesis* that occurred to the resident when he heard the story, was that anuria was *due to* hypovolemia *consecutive on* dehydration *induced by* the vomiting and diarrhea. He accordingly undertook a rehydration test, the negative results of which led him to think that, although dehydration might have *played a role*, it did not suffice to explain the observed renal deficiency.

The patient underwent a first session of dialysis.

Abdominal pain persisted. When X-rays (I.V.U.) brought evidence of a total lack of secretion of the patient's kidneys, the medical team turned to a *second etiological hypothesis*: renal failure would have been the result of a vascular pathology, either a lesion of renal arteries, or a dissection of abdominal aorta. Strong support in favor of the vascular hypothesis was contributed by the presence of numerous 'risk factors' (such as diabetes, hypertension, hyperlipidemia). Arteriography however revealed permeable (though irregular) vessels.

After two more dialysis sessions, the woman felt much better. Anuria however was not relieved, and kaliemia was going up. On May 4 in the evening there was a cardiac arrest. Attempts at resuscitating the patient were unsuccessful. They were abandoned after 45 minutes.

Indeed, the cardiac arrest had very probably been caused by hyperkaliemia, a complication of renal failure. The medical team however could not understand why this had resisted their therapeutic strategy, particularly at a time when the patient's condition had seemed to improve. In fact they felt a little 'cheated' with that death, so that an autopsy was performed and the case was thoroughly studied.

The autopsy was a disappointment. The report read: "no anatomical substratum to the decease". It then mentioned a variety of miscellaneous renal and vascular lesions, no one of which appeared serious enough to have induced death.

The hypothetical causal sequence of events leading to death was so reconstructed:

diabetes ↘ dehydration ↘
hypertension → chronic renal impairment → acute renal failure → hyperkaliemia → cardiac arrest → death
arteriosclerosis ↗

Each event (or state) in the sequence conceivably made the occurrence of the following more probable. Each may also be viewed as a disastrous complication by no means implied by the previous condition.

After discussing the case, clinicians specified as the *main* (underlying) cause on the death certificate: 'acute renal failure', because autopsy had given poor evidence on the seriousness of preexisting renal lesions.

Clinicians were not completely satisfied with the diagnosis, *not* because they assumed there should have been some other underlying condition causing death (the autopsy showed no visceral cancer, for example), but because none of the conditions successively developed by the patient appeared to have been serious enough to explain the fatal outcome. They were nevertheless convinced that 'on the whole' she died of renal complications of a condition rather common in 'developed' countries, including obesity, diabetes, hypertension . . . , of which it is known that they reduce life expectancies.

3.5. The etiological diagnosis goes from observed facts (signs and underlying lesions) to *prior* events causing them (*efficient causes*). A cause of death is a state of the living body, not equivalent to death, but inducing death. The etiological inquiry runs over the decisive steps of the patient's *case history* in the *inverse* chronological order, going from the terminal event to a state or event considered (more or less arbitrarily) as *initial*, and from which the sequence of events leading to death can then be unfolded in the *direct* chronological order.

Thus, in the given example:

– the fact of death having been established, what is its 'direct' cause?–cardiac failure. Note that not all cardiac failures are fatal, otherwise the medical team would not have tried resuscitating the patient.

– given the cardiac failure, what caused it? The patient had been diagnosed to have hyperkaliemia the day before she died (one had even tried correcting it), and hyperkaliemia is known to be toxic for the heart. On the other hand, pathologists had found a badly impaired heart anyhow (coronary arteriosclerosis, dilated left ventricle) without which the potassium overload might have been better tolerated ... Let us admit that the patient died of an hyperkaliemic cardiac failure.

– going back one more step: hyperkaliemia is a frequent complication of renal insufficiency, etc ...

Each step of the diagnostic search can be described in Bayesian terms. Given the effect, its most probable cause is supposed to be the condition which contributed most to its production. The problem in a Bayesian perspective is to see how the set of possible causes is constituted. Medical treatises enumerate all possible causes of pathological conditions, usually ranking them by decreasing frequencies. There also exists the International List of causes of death. These certainly are part of the epistemic background. It would however not make any sense to incriminate a condition of which it had not been ascertained that the patient had it. The etiological diagnosis therefore is dependent on the positive diagnosis of the various lesions and conditions the patient had developed, *and* on the estimate of how severe or serious they were in the particular case (i.e. how likely they were to induce further complications). Medical experience and common statistical knowledge give an estimate of the seriousness (lethality) of various conditions, but this has to be re-estimated for the given case on the basis of the patient's history and symptoms.

3.6. Let us call D_i the individual's death and Z its alleged immediate cause (hyperkaliemic cardiac failure). The probability of Z causing D_i is a function of the probability of Z actually taking place (positive diagnosis of potassium intoxication of the heart) and the probability of death arising from Z, versus the probability of death occurring 'anyhow':

$$P(Z|D_i) = \frac{P(D_i|Z) \cdot P(Z)}{P(D_i|Z) \cdot P(Z) + P(D_i|\neg Z) \cdot P(\neg Z)}$$

Z is considered the cause in case

$$P(Z|D_i) \geqslant P(\neg Z|D_i).$$

From the immediate cause of death, taken as an effect, the same procedure is followed to pick the cause immediately prior to it, say Y (acute renal failure), and so on.

Of course the notion of *immediate* anteriority is relative to a certain level of explanation. Between two consecutive steps in the causal sequence at some (macroscopic) level it is always possible to interpolate a series of intermediate steps at some (microscopic) level, and no *a priori* limit can be assigned to such a process. Whatever the level, the method seems always to regress from effect to cause as far up as possible. Then one may proceed to tell the 'causal history' of the case and unfold the sequence in the direct chronological order, after the causally relevant events have been selected through the Bayesian procedure.

The difficult step in the case cited above was to decide what the cause of acute renal failure had been, because the pathologists had ascertained the existence (not the seriousness) of a chronic nephritis (say X_1), and they had little evidence of an (in itself serious) disease of renal arteries (say X_2), while other causes such as dehydration, say $\neg(X_1 \cup X_2)$, could not be neglected. Making

$$P(Y) = P(Y|X_1) \cdot P(X_1) + P(Y|X_2) \cdot P(X_2) + P(Y|\neg X_1 \cap \neg X_2) \cdot P(\neg X_1 \cap \neg X_2)$$

what had to be weighted was

$$P(X_1|Y) = \frac{P(Y|X_1) \cdot P(X_1)}{P(Y)}, \quad P(X_2|Y) = \frac{P(Y|X_2) \cdot P(X_2)}{P(Y)},$$

and

$$P(\neg X_1 \cap \neg X_2|Y) = \frac{P(Y|\neg X_1 \cap \neg X_2) \cdot P(\neg X_1 \cap \neg X_2)}{P(Y)}.$$

The clinicians were hesitant because, as a matter of fact:

$$P(X_1) > P(X_2) \quad \text{while} \quad P(Y|X_2) > P(Y|X_1)$$

So, even though the causal chain went farther back in time, they officially stopped at Y and designated it as *THE cause* when filling out the standard certificate. Y is neither *initial* nor *ultimate*, nor the cause that gives death the higher probability, for

$$P(D_i|Z) > P(D_i|Y)$$

Thus there is both an element of *chance* in the causal sequence (notion of 'complication': renal failure did not *imply* death), and an element of *arbitrariness* in the choice of the 'main' cause. It seems to me that both aspects are essential in causal analysis. Sigmund Freud wrote: "*our habit is to lay stress on the fortuitous causation of the death*:

accident, disease, infection, advanced age; in this way we betray an effort to reduce death from a necessity to a chance event" [Thoughts for the times on war and death, 1915, §2]. Just delete 'advanced age' and we obtain an accurate description of what causal explanations actually do.

3.7. Statistically speaking, a cause of death is that which shortens the mean duration of life, and its *influence* is measured by the gain in life expectancy resulting from its suppression (see 1, smallpox).

In order for a *medical* cause of death to have a statistical existence, it has to be diagnosed and recognized as lethal by the medical profession, so that the measure of the *objective* influence on mortality of factors like smallpox, phtisis, cancer, etc ... is rooted in the many individual judgments made by medical practitioners on the role of such factors in the deaths of their patients.

Rather than emphasizing the *subjectivity* of such judgments I prefer to insist on their inevitable *particularity*, in the sense that they are made to explain what happened to one individual, the causation of whose death is unique. No two cases of renal failure are identical, even though a generic term is used. A particular explanation has to be made up for each case, and at each step of the Bayesian procedure all the probabilities have to be reassessed on the basis of all available information about the case history.

How arbitrary?

A cardiac failure constitutes an immediate threat of death. Acute renal failure involves a serious risk of death, however not within minutes of its onset. A chronic renal lesion impairs the prospects of survival in the long run. Diabetes, hypertension, ... are 'slow' risk factors. Obesity is compatible with an almost 'normal' life expectancy.

On the other hand obesity exposes one to diabetes, diabetes is a hazard for the kidneys, renal lesions increase the likelihood of acute renal failure, which may induce hyperkaliemia and cardiac arrest.

As one goes farther back in the series of causes, the probability of immediate death (within, say, a day) given the specified cause gets smaller. Why then not designate as *THE cause*, the one which gives death the *maximum probability*, namely the immediate cause (cardiac failure)? Because it is 'immediate' only on the decision to neglect mentioning that it resulted in cerebral anoxia which actually is the 'true' cause of death: for, as long as the brain cells have not been deprived of oxygen for long enough, cardiac failure is not actually mortal (i.e. the causal sequence still may take another turn and not end up in death). Then, why not designate cerebral anoxia as *THE cause*? Because it *is* death itself. What is being looked for is *an efficient cause, not a definition*... An efficient cause is such that the effect always gets a chance not to ensue.

I simplified the situation in implying that risks keep increasing as the causal history develops. The patient might have recovered after being in great danger, she might have lived one more year and died of another illness. Physicians commonly say: "anything may happen". Almost anything.

As the specified cause is a factor with respect to which death is a chance event, its identification is *informative*; all the more informative, it seems, as one goes back further and points to more remote antecedent causes. Saying that death is due to cardiac failure is almost uninformative, for terminal cardiac failure is a fairly trivial fact (actually international recommendations to certifying doctors include the advice *not to* mention cardiac failure on death certificates). Now why not try to *maximize information*, and declare that the *prime* cause of this death was: obesity?

It is interesting to establish that obesity cuts short the mean life, *statistically speaking*. It is almost impossible to affirm that obesity caused this individual's death. One would be tempted to say that it caused it 'indirectly', *via* diabetes and renal failure. But from the fact that obesity rendered diabetes more probable, and that diabetes increased the risk of renal lesions, it cannot be inferred that obesity caused renal lesions, and a fortiori death. Due to non-transitivity, one cannot go back up too far in the causal sequence, and *THE cause* must be picked in the 'middle' zone. Within that zone, there is some arbitrariness in the choice (as confirmed by variations of international doctrines on the point). In the given example, it seems that clinicians stopped just below what was felt as a weaker causal link (for the *post mortem* examination had not definitely established the existence of serious anatomical renal damage).

3.8. It may be said that renal failure (*efficient cause*) *explains* this death, insofar as it significantly increased its probability of occurrence. It explains it the better, as one is apt to 'unfold' the intermediate series of events taking place between the specified cause and the terminal outcome. A complete unfolding would amount to a minute reconstruction of a portion of the individual's history.

Physicians often feel that a patient who died of a condition (say, renal failure) *might not* have died of it, or even (when the patient is young) that he *should not* have died of it. The death of *A* is said to have been *caused by* renal failure, yet *B* will survive a more serious renal failure and live twenty more years. This does not mean that renal failure is not the 'true' cause of *A*'s death: one knew the patient and his condition, and that he did not die of cancer, or septicemia, or smallpox. It does not mean that by inquiring into more minute details of the causal sequence and unfolding some of the intermediate events one will make *A*'s death look more 'necessary': on the contrary – due to non-transitivity – it is not infrequent that in multiplying the number of 'intermediate' steps between the specified 'initial' cause and death one makes the whole sequence look *even more contingent*.

"*Here, as in the rest of life, the more you understand, the less confident you are*", said Condorcet [*Oeuvres*, t. I, p. XIII].

But the (theoretical) possibility of unfolding the detailed series of events compensates for the fortuitous character of the causal sequence by attesting to the continuity, in his personal identity, of the 'carrier' of the successive accidental complications. Entities such as general causal sequences do not exist; there are destinies of individual beings.

3.9. The foregoing suggests some comments on the use of counterfactual conditionals in medical contexts.

"If hyperkaliemia had been corrected, she would not have died." –

"If she had not had nephritis, she would not have developed acute renal failure." – "*If not the cause, then not the effect.*"

It would be preposterous to assume that such sentences (often uttered by doctors) should be logically analyzed as

$$\neg C \supset \neg E$$

for then physicians would be saying that the effect (death) implies the cause (renal failure), knowing that it does not.

The suggested probabilistic approach would allow an alternative interpretation of such counterfactual conditionals.

We are all due to die after all. Life tables assign an *annual risk* for each age (and the corresponding life expectancy). Insurance companies know how to refine the assessment of the risk for a given person.

A *cause of death* is merely a cause of *premature death*, i.e. an event the occurrence of which suddenly *aggravates* the risk. Its non-occurrence does not eliminate the prior risk.

If hyperkaliemia had been corrected, that woman would have had a lesser chance to develop instant heart failure, but she would have been left with the risk involved in her renal deficiency, which was far from negligible. *If she had been rehydrated earlier*, she might have recovered from her acute renal failure, but she would not have been cured of her diabetic-hypertensive-etc ... condition. One never gains or loses more than a little time.

If a cause of death is a factor that 'makes a difference', and if the *influence* of some cause C may be measured by the difference between the probability of this death, given C, and the probability of this death from all other causes,

$$P(D_i|C) - P(D_i) = \epsilon$$

then the counterfactual conditional should be understood as stipulating that, without the cause occurring, the probability of the effect would not have been different from what it was otherwise

$$P(D_i|\neg C) \simeq P(D_i)$$

This may also give an account of the fact that the deaths of old people are often felt to be more difficult to refer to specific causes than the deaths of younger people: the alleviation of one condition does not make the same 'difference' whether one is young or old. "*Had he not developed X, he would not have died*" ... such a statement is uttered with far more conviction when the deceased is a young adult than when he is a debilitated senile person, because the aged person has a higher 'absolute' chance of death anyhow.

It has sometimes been deplored that physicians tend to mention five, six, seven ... causes when certifying the death of old people: this tendency suggests that they cannot admit that death would have been significantly retarded, should only one of the mentioned conditions have been cured. However, he who enumerates a large number of causes implies finally that the old person would not have died, should he not have been ... old. Which refers the explanation of death back to its formal cause, whereas statistical offices obviously expect death certificates to be informative of efficient (preventable) causes.

* * *

4.0. I shall conclude with a few reflections on the *certainty of diagnoses*, on the possibility to judge *scientifically* of *particular etiologies*, and on the *reasoned arbitrariness* of causal ascriptions.

Instructions issued by the W.H.O. since 1948 for coding the information reported on death certificates expressly stipulate the following:

> Qualifying expressions indicating some doubts as to the accuracy of the diagnosis, such as "apparently", "presumably", "possibly", etc., should be ignored, since entities without such qualification differ only in degree of certainty of the diagnosis.
>
> *Manual of the international statistical classification of diseases, injuries and causes of death*, 6th revision, 1948, W.H.O., Geneva, 1950, vol. I, p. 352.

Consequently, two reports such as: "death was *due to* acute renal failure", and "death was *probably due to* acute renal failure", are coded the same way. Thus, according to international rules, the expression *probable cause* is a redundancy.

Laplace, in fact, ignored questions relative to the *probability of judgments*, when discussing the *probability of events*. This occasionally puzzled commentators. "Why did he not follow the fashion of the times and write about degree of belief?" asked F. N. David [1963, p. 34].

Some of Laplace's contemporaries found it difficult to distinguish the 'objective' probability of events from the 'subjectivity' of probability assignments over events. Condorcet for example: "*not only is there no necessary connection between the probability and the actuality of events, but there is no connection either between the computed probability and the real probability*" [1785, p. LXXXVII]. Buffon went farther: he deemed that any adult in good health disregards the possibility of dying within the next 24 hours, a possibility he estimated around .0001. He concluded that in general, any event the *objective* possibility of which is less than .0001, has a zero probability *for us*, "and by the same token, any event the possibility of which is greater than .0001 gives us a complete moral certainty" [Buffon, 1760, §VIII]. In a letter addressed to Buffon, Daniel Bernoulli disputed over the figure and preferred .00001; he however basically accepted Buffon's perspective.

Along those lines, one might hold that on death certificates, the etiological diagnosis is usually asserted without specifications, whereas the alleged cause is a factor which actually made death *only more probable.* On the other hand, one might also hold that the death of the individual was *actually determined* by some causes, which are identified by the certifier *only with a certain probability.* Is the probability of a *probable cause* imputable to medical ignorance, or to the factual contingency of death on its causes?

In fact Condorcet, after discussing at length Buffon's estimate, suggests that everything reduces to a question of *decision,* rather than of *belief* [cf. 1885, p. 224–9].

When analyzing diagnostic procedures along Bayesian-Laplacean lines, there is no need to impose a distinction between the probability that the alleged disorder actually is the cause of death, and the probability that the etiological diagnosis is correct. The first may be taken to yield a measure of the second.

There is indeed an element of subjectivity in the etiological diagnosis, or rather, as was remarked above, an element of *particularity*; for the diagnosis is a singular probability judgment on a particular event. Even though its epistemic background eventually includes the knowledge of the *mean risk* connected with the disorder, the judgment itself implies the reassessment of the risk for the given case. Now, what is evaluated 'to the best of the doctor's knowledge' is the risk $P(D_i|C)$ inherent to the illness, not to the judgment. And when from there he judges of the probability $P(C|D_i)$ for the cause of this death to be C (rather than any other cause), such a probability *is* a measure of the certainty of diagnosis.

The physician obviously does not *explicitly* give an estimate of the probability; however, by designating a *cause,* he *implicitly* indicates that the link between death and the causal event is *merely probable*: for that is precisely what is meant by a *cause.* He does not need to be committed further and decide in favor of ontological determinism vs. human ignorance (probabilistic guess), or of ontological contingency vs. human intelligence (measure). The rational diagnostic procedure is compatible with either.

4.1. Lacroix said that Moses Mendelssohn may have been the first author to suggest an application of probability to the connection between effects and causes, in his *Treatise on evidence* which won the prize of the Academy of Berlin in 1763. The topic proposed by the

Academy for the competition was: 'the mathematical method in philosophy'.[2]

If we have once experienced that two facts *A* and *B* immediately follow each other, three suppositions can occur to us: either *A* has its ground in *B*, or *A* and *B* have their common ground in a third cause *C*, or *A* and *B* are dependent upon separate, i.e. independent causes. In the two first cases, they will have to recur always following each other; in the third,... their coincidence will be the effect of chance: they will be apt to be found separated and remote as well as conjoined...

Attributed to Moses Mendelssohn [1763] by Lacroix, 1833, §103, p. 181–2.

How can probability calculus be applied to non-repeatable causal sequences? Lacroix, speaking of meteorological observations, and quoting Hume, said that the "feeling of belief" follows from the fact that 'positive' signs prevail over 'negative' signs: "the course of reasoning takes the form of a sort of calculus, the result of which acquires more influence on our belief, as judgments and observations are repeated" [1833, §5, p. 8]. There is no doubt that learning how to judge of an illness's seriousness is comparable to learning how to judge of the coming of a rainstorm. However, rather than sticking to the idea of a 'belief' acquired through repetition (would the hundredth diagnosis of smallpox be more certain than the tenth?), the important point is to determine the part attributable to *calculation* and the part attributable to the *estimation of particular probabilities* for the case.

Newton, at the end of his *Opticks*, wrote that "the investigation of difficult things by the Method of Analysis, ought ever to precede the Method of Composition" [third edition, 1721, p. 380]. The analysis consists both in drawing general conclusions from particular experiments (by induction) and inferring causes from the observation of effects; "and the synthesis consists in assuming the causes discover'd, and establish'd as principles, and by them explaining the phenomena proceeding from them, and proving the explanations" [ibid, p. 380–1].

Causes indeed explain their effects: the renal failure explained the patient's death, insofar as it significantly raised its probability of occurrence. This is not to say that causes are 'principles'. According to the classical paradigm of causal explanation, the search for causes merges with inductive procedures, and causal explanations amount to deductions of particular effects-consequences from universal laws. The two aspects are obviously dissociable. The Bayesian procedure which allows regression from (particular) effects to (particular) causes is deductive throughout, and the expository process which restitutes the causal sequence is not a deduction of effects from causes, but a way of setting forth, as rationally as possible, a *history* from which the particularities and contingencies have not eventually been eliminated. Death is the terminal event in the sequence, not the conclusion of an argument. This presents us with a paradigm of causal explanation different from the classical (deductive-nomological) pattern, and compatible with M. Foucault's suggestion [cf. 1963, preface] that 19th century medicine initiated a way of talking scientifically about individuals while respecting their uniqueness. Etiological diagnoses do *not* assert cause-effect relationships *on the basis of repetition*, for the causal sequence they establish is irreversible and cannot be reproduced, and another explanation has to be made up for each case. Yet (in spite of the uniqueness) the procedure described may be considered *scientific*: holding that the most probable cause is that which gives the effect the higher probability of occurrence, is setting a criterion against which causal hypotheses can be tested.

4.2. The apparent arbitrariness in the designation of the cause does not so much pertain to the particularity or to variations in 'degrees of belief', as to a *decision* to maximize information, in particular 'useful' information, usually with a view to stressing the importance of preventable or curable causes. This does not conflict with the results of the Bayesian procedure, but forces it into a calculus of 'mathematical expectation'.

The German physician Johannes Orth emphatically remarked that to the question: "what was THE cause of death?" very different answers may be appropriate, according to whether the question is asked by a pathologist, a statistician working for an office of public health, a member of the family or a lawyer, etc . . . "Any event in the causal chain is A cause," he said, "and may be a candidate for being THE cause" [1908, p. 486].

The best candidate is not evidently the most 'causative' according to the probabilitistic criterion. If it were (see 3.7), THE cause of death would be the terminal coma, or even death itself, and causal explanation would collapse into definition. That is conceivable, insofar as defining death involves stating its formal and/or material causes. However, the possibility of *measuring* the influence of deadly hazards in terms of the extent to which they cut short life expectancies, or to which they raise the probability of impending death, could make sense only within a community that was becoming aware of the fact that the 'law of human mortality' is not immutable, and that the *power* of man over death can be exerted not only in provoking it, but also in retarding it. In that perspective, a cause of death (efficient cause) is something which one should *not* die of. Hence the conceptual effort to reduce death "from a necessity into a chance event", which may well be at the core of causal explanation.

University of Paris XII

ACKNOWLEDGEMENTS

Some of the ideas developed in this paper were first discussed at Stanford during the summer of 1976 with P. Suppes, to whom I owe much. I subsequently benefited from I. Hacking's comments. The paper itself is part of a larger work inquiring into the causation of death: the initial project was to try and define what is meant by a '*cause*' when talking about '*causes of death*'. The original paper was written in French. It has been translated by the author, including all quotations (a few quotations of Laplace, purposely left in the French, are paraphrased in the text). I wish to express my gratitude to E. Adams (Berkeley) and J. Vickers (Claremont) for helping with the translation and making useful remarks.

NOTES

[1] The number that will catch smallpox, $1/n\ s\ dx$, is here identified with the negative of augmentation, ds, in the number, s, of surviving individuals having had smallpox. This identification is questionable [E. Adams].

[2] The German original version of the text was communicated to me by Ian Hacking. There is a Latin translation of it in the 1764 edition, together with a summary in French. I am not aware of an English translation. Lacroix quotes from Degérando's *Histoire comparée des systèmes de philosophie*, which (generously, it seems) grants M. Mendelssohn intuitions on the use of probabilistic notions in causal analysis.

REFERENCES

Bernoulli James [1713, posth.] *Ars conjectandi*, opus posthumum accedit tractatus de seriebus infinitis et epistola gallica scripta de ludo pilae reticularis (Lettre à un amy sur les parties du jeu de Paume, numérotée à part p. 1–35). Basel.

Newton, Isaac [1721] *Opticks*: or, a treatise of the reflections, refractions, inflections and colours of light (the third edition, corrected). London (first edition, 1675).

Deparcieux [1746] *Essai sur les probabilités de la durée de la vie humaine*, d'où l'on déduit la manière de déterminer les rentes viagères, tant simples qu'en tontines, précédé d'une courte explication sur les rentes à terme, ou annuités, et accompagné d'un grand nombre de Tables. Fr. Guérin.

de Moivre A. [1756] *The doctrine of chances*: or, a method of calculating the probabilities of events in play. Followed by: 'A treatise of annuities on lives' (first ed. 1724). London, third edition (previous editions 1711, 1718). Photographic reprint: U.S.A., Chelsea Publishing Company, 1967.

Bernoulli, Daniel [1760] 'Essai d'une nouvelle analyse de la mortalité causée par la petite vérole, et des avantages de l'inoculation pour la prévenir', in: *Mémoires de mathématiques et de physique tirés des registres de l'Académie Royale des Sciences*, de l'année 1760. Hist. de l'Acad... Paris, 1766, 1–45.

Euler, Leonard [1760] 'Recherches générales sur la mortalité et la multiplication du genre humain', suivi de: 'Sur les rentes viagères', in: Hist. de l'Acad... Berlin, 1767, 144–164.

Buffon, Georges Louis Leclerc de [1777] 'Essai d'arithmétique morale', 46–148. 'Des probabilités de la durée de la vie', 149–323, in: *Histoire naturelle, générale et particulière*, servant de suite à l'histoire naturelle de l'homme, supplément, tome quatrième. Fr., Imprimerie Royale.

d'Alembert, Jean le Rond [1761] 'Sur l'application du calcul des probabilités à l'inoculation de la petite vérole', in: *Opuscules*, vol. 2, 26–95.

Mendelssohn, Moses [1763] 'Abhandlung über die Evidenz in den metaphysischen Wissenschaften', in: Moses Mendelssohn's *Gesammelte Schriften*, Leipzig, 1843, Bd. II, 1–64. French abstract and Latin translation in: *Dissertation*, qui a remporté le prix proposé par l'Académie Royale de Sciences et Belles Lettres de Prusse, sur la nature, les espèces et les degrés de l'évidence. Berlin, 1764.

Bayes, Thomas [1763] 'An essay towards solving a problem in the doctrine of chances,... communicated by Mr. Price', in: *Philosophical transactions*, 1764, vol. LIII, 370–418.

Condorcet [1783] 'Mémoire sur le calcul des probabilités, quatrième partie: réflexions sur la méthode de déterminer la probabilité des événements futurs, d'après l'observation des événements passés', in: *Histoire de l'Académie Royale des Sciences*, Paris, 1786, 539–553.

Condorcet [1785] *Essai sur l'application de l'analyse à la probabilité des décisions rendues à la pluralité des voix*. Fr., Imprimerie Royale; photographic reprint, U.S.A., Chelsea Publishing Company, 1972.

Trembley, Jean [1796] 'Recherches sur la mortalité de la petite vérole', in: *Mémoires de l'Académie*... Berlin, 1799, 17–38.

Bichat, Xavier F. [1800] *Recherches physiologiques sur la vie et la mort*. Fr., Masson, 1866.

Duvillard, E. E. [1806] *Analyse et tableaux de l'influence de la petite vérole sur la mortalité à chaque âge, et de celle qu'un préservatif tel que la vaccine peut avoir sur la population et la longévité*. Paris, Imprimerie impériale.

Laplace, Pierre Simon de [1774] 'Mémoire sur la probabilité des causes par les événements', t. VIII, 27–65 (first published in: Mem. Aca... Paris, 1774).

[1778] 'Mémoire sur les probabilités', t. IX, 383–485 (first published in: Mem. Aca ... Paris, 1781).

[1795] 'Leçons de mathématiques, ... dixième séance: sur les probabilités', t. XIV, 146–177.

[1812] 'Théorie analytique des probabilités', t. VII.

[1819] 'Essai philosophique sur les probabilités', t. VII, in: Introduction. *Oeuvres Complètes*, publiées sous les auspices de l'Académie des Sciences, Paris, Gauthier-Villars, 1878–1912, 14 volumes.

Lacroix, S. F. [1833] *Traité élémentaire du calcul des probabilités.* Fr., Bachelier, third edition revised (first edition 1816).

Villermé, L. R. [1833] 'Des épidémies sous le rapport de l'hygiène publique, de la statistique médicale et de l'économie politique', in: *Annales d'hygiène publique et de médecine légale*, Paris, t. IX, part 1, 5–58.

Quételet, A. [1835] *Sur l'homme et le développement de ses facultés*, ou essai de physique sociale. Belg., publ. Paris, Bachelier, 2 vol.

Bernoulli, Christoph [1841] *Handbuch der Populationistik*, oder der Völker- und Menschenkunde nach statistischen Erlebnissen. Ulm, 3 vol.

Cournot, M. A. A. [1843] *Exposition de la théorie des chances et des probabilités.* Fr., Hachette.

Mill, John Stuart [1843] *System of logic* ratiocinative and inductive being a connected view of the principles of evidence and the methods of scientific investigation. G. B., Longman, new impression 1906.

'La mortalité à Paris en 1870', a review of a study by Dr. Vacher which appeared in the *Gazette Médicale*; followed by an analysis of cause of death returns in Paris for the month of January, 1871 (bombardments), by Ch. Daremberg. The third paper comes under the heading: 'documents pour servir a l'histoire du mouvement scientifique pendant le siège de Paris'. In: *Gazette des hôpitaux civils et militaires* (La Lancette française). January 24: 10, 39, June 1: 65, 259, June 3: 66, 263–4.

Todhunter, I. [1865] *A history of the mathematical theory of probability from the time of Pascal to that of Laplace.* G. B.; textually unaltered reprint, U.S.A., Chelsea Publishing Company, 1965.

Farr, William [1885, posthumous] *Vital statistics*: a memorial volume ... edited for the Sanitary Institute of Great Britain by Noël Humphreys. London.

Pearson, Karl [1897] 'The chances of death', in: *The chances of death and other studies in evolution*, vol. I, 1–41. G.B., Arnold.

Orth, Johannes [1908] 'Was ist Todesursache?', in: *Berliner Klinische Wochenschrift*, 45, 10, 485–490.

Orth, Johannes, *Manual of the International Statistical Classification of Diseases, Injuries and Causes of Death*, sixth revision of the international lists of diseases and causes of death, adopted 1948. Bulletin of the World Health Organization, supplement I, W. H. O., Geneva, 1948, 2 vol.

Aubenque, M. [1956] 'Statistique et causalité, in: *Journal de la société de statistique de Paris*, juil-sept 1956, 166–181.

David, F. N. [1963] 'Some notes on Laplace', in: *Proceedings* of an International Research Seminar, Bernoulli-Bayes-Laplace anniversary volume, University of California, Berkeley, 30–44. U.S.A., Springer-Verlag.

Hacking, Ian [1965] *Logic of statistical inference.* G. B., Cambridge University Press, repr. 1976.

Suppes, Patrick [1970] 'A probabilistic theory of causality', in: *Acta philosophica fennica,* fasc. XXIV, North Holland Publishing Company.

Hacking, Ian [1975] *The emergence of probability,* a philosophical study of early ideas about probability, induction and statistical inference. G. B., Cambridge University Press.

Suppes, Patrick [1976] 'Hume and Laplace on causality'. U.S.A., xeroxed.

Suppes, Patrick [1977] 'The logic of clinical judgment: Bayesian and other approaches'. 5th transdisciplinary symposium on philosophy and medicine, U.C.L.A., Apr. 14–16, 1977. In: Engelhardt, H. T. Jr., Spicker, S. F., Towers, B. (eds.) [1979] *Clinical judgment: a critical appraisal.* Dordrecht, Reidel, 145–159.

Manual of the international statistical classification of diseases, injuries and causes of death, based on the recommendations of the 9th revision conference, 1975, and adopted by the 29th World Health Assembly. W. H. O., Geneva, 1977, 2 volumes.

Age. Curt.	Per-sons.	Age. Curt.	Per-sons	Age Curt.	Per-sons	Age. Curt.	Per-sons	Age. Curt.	Per-sons	Age. Curt.	Per-sons
1	1000	8	680	15	628	22	586	29	539	36	481
2	855	9	670	16	622	23	579	30	531	37	472
3	798	10	661	17	616	24	573	31	523	38	463
4	760	11	653	18	610	25	567	32	515	39	454
5	732	12	646	19	604	26	560	33	507	40	445
6	710	13	640	20	598	27	553	34	499	41	436
7	692	14	634	21	592	28	546	35	490	42	427
Age Curt	**Per-sons.**	**Age. Curt.**	**Per-sons**	**Age Curt.**	**Per-sons**	**Age Curt**	**Per-sons**	**Age. Curt.**	**Per-sons**	**Age. Curt.**	**Per-sons**
43	417	50	346	57	272	64	202	71	131	78	58
44	407	51	335	58	262	65	192	72	120	79	49
45	397	52	324	59	252	66	182	73	109	80	41
46	387	53	313	60	242	67	172	74	98	81	34
47	377	54	302	61	232	68	162	75	88	82	28
48	367	55	292	62	222	69	152	76	78	83	23
49	357	56	282	63	212	70	142	77	68	84	20

Age.	Persons.
7	5547
14	4584
21	4270
28	3964
35	3604
42	3178
49	2709
56	2194
63	1694
70	1204
77	692
84	253
100	107
	34000
	Sum Total.

1. **Edmund Halley [1693] 'An estimate of the degrees of the mortality of mankind' . . . *Philosophical transactions*, XVII, p. 600.**

The Probabilities of human Life, according to different Authors.

Table I, by Dr. *Halley*.

Age	Living.	Age	Living.	Age	Living.	Age	Living.	Age	Living.	Age	Living.
1	1000	16	622	31	523	46	387	61	232	76	78
2	855	17	616	32	515	47	377	62	222	77	68
3	798	18	610	33	507	48	367	63	212	78	58
4	760	19	604	34	499	49	357	64	202	79	*49
5	732	20	598	35	*490	50	*346	65	192	80	41
6	710	21	592	36	481	51	335	66	182	81	34
7	692	22	586	37	472	52	324	67	172	82	28
8	680	23	580	38	463	53	313	68	162	83	23
9	670	24	574	39	454	54	302	69	152	84	19
10	661	25	*567	40	445	55	*292	70	142	*	*
11	653	26	560	41	436	56	282	71	*131		
12	646	27	553	42	427	57	272	72	120		
13	*640	28	546	43	*417	58	262	73	109		
14	634	29	539	44	407	59	252	74	98		
15	628	30	*531	45	397	60	242	75	*88		

Table III. by M. de *Parcieux*.

Age	Living	Age	Living	Age	Living.	Age	Living.	Age	Living.
1	****	21	806	41	650	61	450	81	101
2	****	22	798	42	643	62	437	82	85
3	1000	23	790	43	636	63	423	83	71
4	970	24	782	44	629	64	409	84	59
5	948	25	774	45	622	65	395	85	48
6	930	26	766	46	615	66	380	86	38
7	915	27	758	47	607	67	364	87	29
8	902	28	750	48	599	68	347	88	22
9	890	29	742	49	590	69	329	89	16
10	880	30	734	50	581	70	310	90	11
11	872	31	726	51	571	71	291	91	7
12	866	32	718	52	560	72	271	92	4
13	860	33	710	53	549	73	251	93	2
14	854	34	702	54	538	74	231	94	1
15	848	35	694	55	526	75	211	95	0
16	842	36	686	56	514	76	192	96	*
17	835	37	678	57	502	77	173	97	*
18	828	38	671	58	489	78	154	98	
19	821	39	664	59	476	79	136	99	
20	814	40	657	60	463	80	118	100	

Table II. by M. *Kersseboom*.

Age	Living.	Age	Living.	Age	Living.	Age	Living.	Age	Living.	Age	Living.	Age	Living.
0	1400												
1	1125	16	849	31	699	46	550	61	360	76	160	91	7
2	1075	17	842	32	687	47	540	62	356	77	145	92	5
3	1030	18	835	33	675	48	530	63	343	78	130	93	3
4	993	19	826	34	665	49	518	64	329	79	115	94	2
5	964	20	817	35	655	50	507	65	315	80	100	95	1
6	947	21	808	36	645	51	495	66	301	81	87	96	0.6
7	930	22	800	37	635	52	482	67	287	82	75	97	0.5
8	913	23	792	38	625	53	470	68	273	83	64	98	0.4
9	904	24	783	39	615	54	458	99	259	84	55	99	0.2
10	895	25	772	40	605	55	446	70	245	85	45	100	0.0
11	886	26	760	41	596	56	434	71	231	86	36	*	
12	878	27	747	42	587	57	421	72	217	87	28		
13	870	28	735	43	578	58	408	73	203	88	21		
14	863	29	723	44	569	59	395	74	189	89	15		
15	856	30	711	45	560	60	382	75	175	90	10		

Y y Table III.

Table IV. by Messieurs *Smart* and *Simpson*.

Age	Living.	Age	Living.	Age	Living.	Age	Living.	Age	Living.
1	1280 } 870 }	17	480	33	358	49	212	65	99
2	700	18	474	34	349	50	204	66	93
3	635	19	468	35	340	51	196	67	87
4	600	20	462	36	331	52	188	68	81
5	580	21	455	37	322	53	180	69	75
6	564	22	448	38	313	54	172	70	69
7	551	23	441	39	304	55	165	71	64
8	541	24	434	40	294	56	158	72	59
9	532	25	426	41	284	57	151	73	54
10	524	26	418	42	274	58	144	74	49
11	517	27	410	43	264	59	137	75	45
12	510	28	402	44	255	60	130	76	41
13	504	29	394	45	246	61	12	77	38
14	498	30	385	46	237	62	117	78	35
15	192	31	376	47	228	63	111	79	32
16	486	32	367	48	220	64	105	80	29

2. **A. de Moivre. *A treatise of annuities on lives*, 2nd ed., 1756, Appendix VII.**

TABLE I.

Ages par années.	Survivans selon M. Halley.	N'ayant pas eu la pet. vérole	Ayant eu la pet. vérol.	Prenant la pet. vérole pendant ch. année.	Morts de la pet. vérole pendant chaq. année.	Somme des morts de la pet. vérole.	Morts par d'autres maladies pend. chaq. année.
0	1300	1300	0				
1	1000	896	104	137	17,1	17,1	283
2	855	685	170	99	12,4	29,5	133
3	798	571	227	78	9,7	39,2	47
4	760	485	275	66	8,3	47,5	30
5	732	416	316	56	7,0	54,5	21
6	710	359	351	48	6,0	60,5	16
7	692	311	381	42	5,2	65,7	12,8
8	680	272	408	36	4,5	70,2	7,5
9	670	237	433	32	4,0	74,2	6
10	661	208	453	28	3,5	77,7	5,5
11	653	182	471	24,4	3,0	80,7	5
12	646	160	486	21,4	2,7	83,4	4,3
13	640	140	500	18,7	2,3	85,7	3,7
14	634	123	511	16,6	2,1	87,8	3,9
15	628	108	520	14,4	1,8	89,6	4,2
16	622	94	528	12,6	1,6	91,2	4,4
17	616	83	533	11,0	1,4	92,6	4,6
18	610	72	538	9,7	1,2	93,8	4,8
19	604	63	541	8,4	1,0	94,8	5
20	598	56	542	7,4	0,9	95,7	5,1
21	592	48,5	543	6,5	0,8	96,5	5,2
22	586	42,5	543	5,6	0,7	97,2	5,3
23	579	37	542	5,0	0,6	97,8	6,4
24	572	32,4	540	4,4	0,5	98,3	6,5

3. Daniel Bernoulli [1760] 'Essai d'une nouvelle analyse de la mortalité causée par la petite vérole, et des avantages de l'inoculation pour la prévenir', *Mémoires de mathématiques et de physique tirés des registres de l'Académie royale des sciences*, Paris, 1766, p. 44.

TABLE II.

AGES par années.	État naturel & variolique.	ÉTAT non-varioliq.	Différ. ou gains.	AGES par années.	État naturel & variolique.	ÉTAT non-varioliq.	Différ. ou gains.
0	1300	1300	0	13	640	741,1	74,1
1	1000	1017,1	17,1	14	634	709,7	75,7
2	855	881,8	26,8	15	628	705,0	77,0
3	798	833,3	35,3	16	622	700,1	78,1
4	760	802,0	42,0	17	616	695,0	79,0
5	732	779,8	47,8	18	610	689,6	79,6
6	710	762,8	52,8	19	604	684,0	80,0
7	692	749,1	57,2	20	598	678,2	80,2
8	680	740,9	60,9	21	592	672,3	80,3
9	670	734,4	64,4	22	586	666,3	80,3
10	661	728,4	67,4	23	579	659,0	80,0
11	653	722,9	69,9	24	572	651,7	79,7
12	646	718,2	72,2	25	565	644,3	79,3

Cette Table fait voir d'un coup d'œil, combien sur 1300 enfans, supposés nés en même temps, il en resteroit de vivans d'année en année jusqu'à l'âge de vingt-cinq ans, en les supposant tous sujets à la petite vérole; & combien il en resteroit s'ils étoient tous exempts de cette maladie, avec la comparaison & la différence des deux états.

4. Daniel Bernoulli [1760] 'Essai d'une nouvelle analyse de la mortalité causée par la petite vérole, et des avantages de l'inoculation pour la prévenir'. *Mémoires de mathématiques et de physique tirés des registres de l'Académie royale des sciences*, Paris, 1766, p. 45.

Années	*n*	*m*	Années	*n*	*m*	Annécs	*m*
1	7	6	19	5,8	60	de 0 à 5 ans incl.	6,3
2	6	6	20	5,8	74	6 - 10	29
3	5,7	6	21	5,8	74	11 - 15	100
4	5,5	5,5	22	5,8	74	16 - 20	65
5	5,5	8	23	5,8	74	21 - 25	90
6	5,5	11	24	5,8	93	26 - 35	67
7	5,5	14	25	5,8	133		
8	5,5	17	26	5,8	133		
9	5,5	30	27	5,8	112		
10	5,5	72	28	5,8	92		
11	5,5	120	29	5,8	78		
12	5,5	112	30	5,8	63		
13	5,5	100	31	5,8	52		
14	5,65	90	32	5,8	44		
15	5,7	80	33	5,8	37		
16	5,7	74	34	5,8	31		
17	5,7	67	35	5,8	26		
18	5;7	63					

5. Jean Trembley [1796] 'Recherches sur la mortalité de la petite vérole', *Mémoires de l'Académie Royale*..., Berlin, 1799, p. 34.

TABLE Ire

Loi de Mortalité (en France), dans l'état naturel.

AGES.	SOMME DES VIVANS.	VIVANS de chaque âge.	MORTS de chaque âge dans l'année.	SUR COMBIEN il en meurt un par année.	VIE moyenne.	VIE probable.
x	$S.y$	y	Δy	$\frac{y}{\Delta y}$	$\frac{S.y}{y} - \frac{1}{2}$	$t \parallel \frac{y_x}{2} = y_{x+t}$
ans.					ans.	ans.
0.	29,2631921.	1,0000000.	0,2324753.	1 sur 4,301.	28,7632.	20,3755.
1.	28,2631921.	0,7675247.	0,0956906.	8,021.	36,3499.	36,9332.
2.	27,4956674.	0,6718341.	0,0471657.	14,244.	40,4253.	42,7424.
3.	26,8228333.	0,6246684.	0,0259550.	24,067.	42,4409.	44,4749.
4.	26,1091649.	0,5987134.	0,0155625.	38,472.	43,2591.	45,6988 *max.*
5.	25,6004515.	0,5831509.	0,0101259.	57,590.	43,4002 *max.*	45,5127.
6.	25,0173006.	0,5730250.	0,0071871.	79,726.	43,1583.	45,3636.
7.	24,4442756.	0,5658379.	0,0055933.	101,164.	42,7001.	44,8224.
8.	23,8784377.	0,5602446.	0,0047582.	117,743.	42,1215.	44,1764.
9.	23,3181931.	0,5554864.	0,0043648.	127,268.	41,4780.	43,4749.
10.	22,7627067.	0,5511216.	0,0042334 *min.*	130,184 *max.*	40,8025.	42,7489.
11.	22,2115851.	0,5468882.	0,0042581.	128,435.	40,1155.	42,0143.
12.	21,6646969.	0,5426301.	0,0043751.	124,027.	39,4254.	41,2768.
13.	21,1220668.	0,5382550.	0,0045445.	118,441.	38,7419.	40,5466.
14.	20,5838118.	0,5337105.	0,0047412.	112,569.	38,0674.	39,8268.
15.	20,0501013.	0,5289693.	0,0049490.	106,884.	37,4041.	39,1170.
16.	19,5211320.	0,5240203.	0,0051576.	101,616.	36,7507.	38,4166.
17.	18,9971117.	0,5188627.	0,0053605.	96,794.	36,1130.	37,7289.
18.	18,4782490.	0,5135022.	0,0055532.	92,470.	35,4839.	37,0525.
19.	17,9647468.	0,5079490.	0,0057331.	88,599.	34,8672.	36,3827.
20.	17,4567978.	0,5022159.	0,0058989.	85,137.	34,2596.	35,7235.
21.	16,9545819.	0,4963170.	0,0060496.	82,041.	33,6562.	35,0728.
22.	16,4582649.	0,4902674.	0,0061849.	79,268.	33,0700.	34,4258.
23.	15,9679975.	0,4840825.	0,0063055.	76,771.	32,4861.	33,7868.
24.	15,4839150.	0,4777770.	0,0064110.	74,519.	31,9082.	33,1520.
25.	15,0061380.	0,4713660.	0,0065026.	72,489.	31,3354.	32,5193.
26.	14,5347720.	0,4648634.	0,0065812.	70,635.	30,7668.	31,8919.
27.	14,0699086.	0,4582822.	0,0066476.	68,939.	30,2014.	31,2642.
28.	13,6116264.	0,4516346.	0,0067030.	67,378.	29,6386.	30,6384.
29.	13,1599918.	0,4449316.	0,0067484.	65,933.	29,0776.	30,0154.
30.	12,7150602.	0,4381832.	0,0067854.	64,577.	28,5177.	29,3887.
31.	12,2768770.	0,4313978.	0,0068150.	63,301.	27,9584.	28,7641.
32.	11,8454792.	0,4245828.	0,0068388.	62,084.	27,3991.	28,1388.
33.	11,4208964.	0,4177440.	0,0068581.	60,913.	26,8395.	27,5111.
34.	11,0031524.	0,4108859.	0,0068743.	59,771.	26,2791.	26,8843.
35.	10,5922665.	0,4040116.	0,0068890.	58,632.	25,7177.	26,2546.
36.	10,1882549.	0,3971226.	0,0069035.	57,526.	25,1552.	25,6239.
37.	9,7911323.	0,3902191.	0,0069190.	56,398.	24,5914.	24,9940.
38.	9,4009132.	0,3833001.	0,0069370.	55,254.	24,0263.	24,3601.
39.	9,0176131.	0,3763631.	0,0069589.	54,084.	23,4599.	23,7270.
40.	8,6412500.	0,3694042.	0,0069857.	52,880.	22,8924.	23,0939.
41.	8,2718458.	0,3624185.	0,0070186.	51,637.	22,3240.	22,4591.
42.	7,9094273.	0,3553999.	0,0070584.	50,351.	21,7550.	21,8260.
43.	7,5540274.	0,3483415.	0,0071064.	49,018.	21,1857.	21,1931.
44.	7,2056859.	0,3412351.	0,0071629.	47,639.	20,6165.	20,5610.
45.	6,8644508.	0,3340722.	0,0072289.	46,213.	20,0478.	19,9318.
46.	6,5303786.	0,3268433.	0,0073046.	44,745.	19,4802.	19,3039.
47.	6,2035353.	0,3195387.	0,0073904.	43,237.	18,9140.	18,6794.
48.	5,8839966.	0,3121483.	0,0074864.	41,695.	18,3500.	18,0590.
49.	5,5718483.	0,3046619.	0,0075924.	40,127.	17,7886.	17,4422.
50.	5,2671864.	0,2970695.	0,0077084.	38,538.	17,2305.	16,8308.
51.	4,9701169.	0,2893611.	0,0078337.	36,938.	16,6762.	16,2256.
52.	4,6807558.	0,2815274.	0,0079677.	35,334.	16,1263.	15,6256.
53.	4,3992284.	0,2735597.	0,0081093.	33,734.	15,5814.	15,0331.
54.	4,1256687.	0,2654504.	0,0082575.	32,147.	15,0421.	14,4487.
55.	3,8602183.	0,2571929.	0,0084108.	30,579.	14,5090.	13,8720.

6. E. E. Duvillard [1806] *Analyse et tableaux de l'influence de la petite vérole sur la mortalité à chaque âge, et de celle qu'un préservatif tel que la vaccine peut avoir sur la population et la longévité*, p. 161.

TABLE II.

AGES.	SOMME DES VIVANS — Qui ont eu la petite vérole.	VIVANS — Qui ont eu la petite vérole.	DIFFÉRENCES.	VIE MOYENNE des individus qui ont eu la petite vérole.	Nombre d'années pendant lequel un individu qui n'a pas eu la petite vérole à l'âge *a* existera — sans avoir la petite vérole.	Nombre d'années pendant lequel un individu qui n'a pas eu la petite vérole à l'âge *a* existera — après avoir eu la petite vérole.
x	$Sw = Sy - Sn$	$w = y - n$	Δw	$\dfrac{\frac{w_a}{z_a}\int z\,\partial x}{w_a}$	$\dfrac{\int n\,\partial x}{n_a}$	$\dfrac{\int (h-n)\partial x}{n_a} = \dfrac{\int w\partial x - \frac{w_a}{z_a} z\partial x}{n_a}$
0.	23,3129978.	0,0000000.	+ 0,0736101.	32,2558.	5,4502.	23,3130.
1.	23,3129978.	0,0736101.	0,0427740.	39,9930.	6,6337.	29,3009.
2.	23,2393877.	0,1163841.	0,0406983.	43,1561.	7,1628.	32,6915.
3.	23,1230036.	0,1570824.	0,0395867.	44,3362.	7,4148.	34,3873.
4.	22,9659212.	0,1966691.	0,0371705.	44,6112 *m.*	7,5420.	35,0557.
5.	22,7692521.	0,2338396.	0,0339893.	44,4235.	7,6051.	35,1101 *max.*
6.	22,5354125.	0,2678289.	0,0305225.	43,9830.	7,6321.	34,8044.
7.	22,2675836.	0,2983514.	0,0270533.	43,3996.	7,6376 *m.*	34,2824.
8.	21,9692322.	0,3254047.	0,0237243.	42,7361.	7,6298.	33,6399.
9.	21,6438275.	0,3491290.	0,0206015.	42,0304.	7,6139.	32,9294.
10.	21,2946985.	0,3697305.	0,0177118.	41,3056.	7,5931.	32,1840.
11.	20,9249680.	0,3874423.	0,0150612.	40,5758.	7,5693.	31,4243.
12.	20,5375257.	0,4025035.	0,0126456.	39,8495.	7,5439.	30,6631.
13.	20,1350222.	0,4151491.	0,0104545.	39,1319.	7,5178.	29,9082.
14.	19,7198731.	0,4256036.	0,0084753.	38,4256.	7,4915.	29,1656.
15.	19,2942695.	0,4340789.	0,0066934.	37,7320.	7,4653.	28,4388.
16.	18,8601906.	0,4407723.	0,0050938.	37,0516.	7,4394.	27,7302.
17.	18,4194183.	0,4458661.	0,0036616.	36,3843.	7,4140.	27,0418.
18.	17,9735522.	0,4495277.	0,0023827.	35,7296.	7,3890.	26,3752.
19.	17,5240245.	0,4519104.	0,0012432.	35,0870.	7,3646.	25,7302.
20.	17,0721141.	0,4531536.	0,0002305.	34,4556.	7,3407.	25,1083.
21.	16,6189605.	0,4533841 *max.*	— 0,0006678.	33,8346.	7,3173.	24,5084.
22.	16,1655764.	0,4527163.	0,0014622.	33,2230.	7,2944.	23,9303.
23.	15,7128601.	0,4512541.	0,0021635.	32,6200.	7,2719.	23,3733.
24.	15,2616060.	0,4490906.	0,0027805.	32,0247.	7,2496.	22,8361.
25.	14,8125154.	0,4463101.	0,0033222.	31,4360.	7,2276.	22,3171.
26.	14,3662053.	0,4429879.	0,0037966.	30,8530.	7,2057.	21,8145.
27.	13,9232174.	0,4391913.	0,0042109.	30,2749.	7,1838.	21,3268.
28.	13,4840261.	0,4349804.	0,0045716.	29,7008.	7,1618.	20,8530.
29.	13,0490457.	0,4304088.	0,0048852.	29,1298.	7,1394.	20,3889.
30.	12,6186369.	0,4255236.	0,0051572.	28,5613.	7,1166.	19,9333.
31.	12,1931133.	0,4203664.	0,0053928.			
32.	11,7727469.	0,4149736.	0,0055970.			
33.	11,3577733.	0,4093766.	0,0057743.			
34.	10,9483967.	0,4036023.	0,0059287.			
35.	10,5447944.	0,3976736.	0,0060642.	25,7337.	6,9901.	17,7252.
36.	10,1471208.	0,3916094.	0,0061843.			
37.	9,7555114.	0,3854251.	0,0062921.			
38.	9,3700863.	0,3791330.	0,0063907.			
39.	8,9909533.	0,3727423.	0,0064829.			
40.	8,6182110.	0,3662594.	0,0065710.	22,8973.	6,8261.	15,5011.
41.	8,2519516.	0,3596884.	0,0066576.			
42.	7,8922632.	0,3530308.	0,0067440.			
43.	7,5392324.	0,3462868.	0,0068;27.			
44.	7,1929456.	0,3394541.	0,0069248.			
45.	6,8534915.	0,3325293.	0,0070217.	20,0490.	6,6032.	13,1958.
46.	6,5209622.	0,3255076.	0,0071243.			
47.	6,1954546.	0,3183833.	0,0072337.			
48.	5,8770713.	0,3111496.	0,0073502.			
49.	5,5659217.	0,3037994.	0,0074741.			
50.	5,2621223.	0,2963253.	0,0076056.	17,2307.	6,3048.	10,8453.
51.	4,9657970.	0,2887197.	0,0077445.			
52.	4,6770773.	0,2809752.	0,0078903.			
53.	4,3961021.	0,2730849.	0,0080421.			
54.	4,1230172.	0,2650428.	0,0081994.			
55.	3,8579744.	0,2568434.	0,0083604.	14,5091.	5,9209.	8,5882.

7. Duvillard E. E. [1806] ***Analyse et tableaux de l'influence de la petite vérole sur la mortalité à chaque âge, et de celle qu'un préservatif tel que la vaccine peut avoir sur la population et la longévité***, p. 163.

IAN HACKING

FROM THE EMERGENCE OF PROBABILITY TO THE EROSION OF DETERMINISM

In terms of scholarship, imagination and sheer skill of exposition the papers by Fagot and Schneider are models for us all. I shall address specific remarks to each of them, and to the commentaries by Nový, Kostiouk and Cohen,* but first a few general observations. We are all of us engaged in different enterprises which I regard as entirely collaborative. Schneider is an historian of mathematics. Fagot is qualified as a medical doctor and is at present teaching at an Institute for the history of knowledge and ideas. I am a philosopher who subtitled *The Emergence of Probability* not a "history" but a "philosophical study" of early ideas about probability, induction and statistical inference. We all have our own guild battles to fight. Schneider shows in his paper that he disagrees, on certain points, with his celebrated colleague van der Waerden. Fagot has some difficulty convincing the medical profession that her archaeological enquiries into the causes of death could teach one anything about the idea of a cause of death. And philosophers doubt that my work contributes to the analysis of the concept of probability. I am asked, "you're supposed to work on the logic of statistical inference!"

We three draw on the same data but put different questions. I know that the answers made by Schneider and Fagot to their own questions are crucial to my work, and I hope that some of my answers bear on their research. Schneider asks, when did the probability calculus begin, and why then ? I ask, when did the entire mode of probabilistic conceptualizing begin, and why then ? The questions do not have quite the same answers, but I think neither can be properly answered without the other. His question has a straightforward motivation. Mine is more Byzantine and I shall take a little time to explain it, for this conference is not only about the facts of, in this instance, old probability ideas, but also about how philosophy and history can interrelate in scrutinizing these facts.

A few days ago Dr Fagot kindly allowed me to consult some materials and by chance her notes on my own book were amongst

I. Hintikka, D. Gruender, and E. Agazzi (eds.), Pisa Conference Proceedings, Vol. II, 105–123.

them. Lacking the strength of character not to glance at what she had written, I was pleased with the one sentence summary: "An investigation in the style of Michel Foucault, although a great deal more clear." I could not put it better myself! It is worth saying a few words about this archaeological style that I learned by reading Foucault's fascinating and infuriating works. In *The Emergence of Probability* I claim that there is a coming into being, all at once, of a certain organization of concepts, which persists through a substantial period of time. What emerges is just what Alistair Crombie elsewhere in this conference has been calling the "statistical style," not just a body of knowledge, but a way of thinking, a way of talking. I give a Foucaultian before-and-after picture of this mutation, a picture of before 1650 or so, when there is virtually none of our organization of probability ideas. Then I give a picture of after 1670 or so when (with one exception) the main elements of probability are present.

This emergence is not a matter of particular heros – though I cite de Witt and Graunt and Leibniz and Huygens and Pascal and Wilkins and Petty and so on. It is a matter of a spontaneous emergence of a new style in reasoning about man and God, of describing nations, of existential theology, of merchant adventurers and the methods of science. It has to do with a vast range of new conceptions, such as new practices of trade and the idea of the state (not discussed in my book) and also with the new science of the seventeenth century, with its new canons of evidence (amply discussed in the book).

It is part of my Foucaultian claim that there came into being a new class of sentences, of new things to say, of sentences which for the first time have a "positivity," *i.e.* are "up for grabs" as being possibly true or false. I maintained that all this became possible only with the coming into being of a new kind of evidence. Joseph Agassi claimed yesterday that we should think about changing standards of rationality; in the book I asserted something more specific, that a particular kind of evidence is novel, and accounts for the emergence of probability. Moreover I said that the space of probability ideas – a space of sentences with truth values – comes into being fully fledged. With some exaggerations I maintained that all possible philosophical positions about probability are already represented in the early days, and that we can understand the structure of current philosophical problems about probability by understanding what made possible this concept at all. Thus there is a claim that the preconditions for the possibility of this concept are what determine the incoherences that prompt philosophical discussion today in, for example, the foundations of statistical inference.

Such was my enterprise, manifestly different from and far more obscure than Schneider's, although with some central points in common with Fagot's research. I deliberately caricature my own contributions in this abstract way because my book is perhaps too full of the minutiae of historical fact. When combined with a certain shyness of mine when it comes to expressing general themes, the plethora of details make it difficult for some readers to see what I have been doing. I shall have to apologize to L. J. Cohen, for example, for having written in such a way that he could not see what the book is about. My enterprise demands fidelity to thicket upon thicket of historical detail, But I would not want the shape of the forest to be forgotten. I would far rather have *Emergence* dismissed as preposterous than for its philosophical theses to pass unnoticed.

1. SCHNEIDER'S ORIGINS OF THE PROBABILITY CALCULUS

John North said a couple of days ago that if philosophers are to make claims that bear on the growth of knowledge, then they must be rigorously testable. No one will suppose that my simplistic picture can possibly be correct: human knowledge is too complex a web for that. One point of enunciating a simplistic sketch is to make assertions that might be false. Thus I claimed that a certain idea of probability occurs in any number of sentences written after 1650 or so, but not at all before. Schneider disagrees. There was, he insists, an ancient idea of probability indicated even in certain familiar Aristotelian passages, and more importantly in a sceptical tradition of which Carneades is the best example. What is more, these very passages are referred to in just those seventeenth century texts that constitute what I call the "emergence of probability." So how can I claim that probability is a new invention, and not just one more rediscovery of the Renaissance?

I have two comments on this. One is a dull bit of saving face, but the other is interesting. As to the former I did *not* claim that there had been no idea of probability, ever, and indeed produced a text that implies the existence of a curiously sophisticated probability notion perhaps as early as 400 B.C. My claim was only that there was a sudden eruption of probability lore around 1660, and no such lore in the preceding epoch, and that the way that probability erupted has to do with a transition of Renaissance ideas, chiefly connected with what I call the low sciences. Now it is trivially consistent with that thesis that there should have been some probability talk a millenium earlier, talk which was noted by the more scholarly seventeenth century authors – although no one would contend that merchants such as Graunt or adventurers like Petty were moved, in this spontaneous emergence of probability, by dipping into Carneades. Moreover what is important is not the occurence of a few probability ideas in antique texts but a use for them, a use that spans morals, politics, economics and social affairs, and which engenders a new era of conjecturing on the one hand, and a new mode of representing reality on the other. "I must discuss the concept of sign," (I wrote on p. 17), "a good stoic concept, much debated in for example Sextus Empiricus, and itself referring back to Aristotle. Those ancient sentences are not irrelevant, but Sextux matters to [*Emergence*] not because of what he said but only because of the use that Gassendi, say, made of him" in mid-seventeenth century.

So much for my face-saving response. Shortly before receiving Schneider's paper I learned of another essay about Carneades: Miles' Burnyeat's "Carneades was no probabalist." Burnyeat recalls a long tradition that includes the article on Carneades in Bayle's dictionary, and, in our times, a paper of Roderick Chisholm that compares Carneades to Reichenbach.[2] But, says Burnyeat, "the whole thing is a myth, more precisely, a mistranslation of Carneades' key term *πιθανόν*."

Disputes about translating this word are not new. Burnyeat's originality lies in propounding a new reading of Carneades in which anything like our concept of probabality would be anachronistic. It is not for me to take sides in this philological dispute. Indeed in response to the remark of North's that I mentioned above, I would cite my philosopher's doctrine about the emergence of probability as one which is now seen to have unexpected empirical consequences about what Carneades said. Conversely Burnyeat, although not committing himself to *Emergence*, does observe that the claims made there nicely conspire with his own conjectures. To repeat, I am not saying that Schneider is mistaken, only that there is a real philological dispute about Carneades whose resolution will point in different directions, one favourable to Schneider, one to me.

Schneider has since indicated that it may not matter what Carneades himself intended; it is enough that Leibniz, Bayle and so on report him as having probability ideas. My view is quite the contrary. I claim that because of the emergence of probability in the mid-seventeenth century, these authors read Carneades and his ilk in a certain way, imposing on those writers certain concepts that have newly evolved.

I must add a note of caution about my doctrine. Since the conference I have received a paper by Daniel Garber and Sandy Zabell which takes issues with my claims to the novelty of probability.[3] They adduce a large number of texts from Cicero on, arguing that there is ample evidence of what we call probability. Evidently I cannot discuss this careful work here, but I do give notice that just as I have found an unexpected ally in Burnyeat, so Schneider will find welcome confirmation in Garber and Zabell.

Schneider's paper has a good many extremely interesting observations which, as he says, do not conflict with my analysis. I particularly admire his conjecture about early occurrences of the division problem ("the problem of points," of how to divide the stakes in a game of chance that has not been completed, but in which one party has the advantage over the other). He says that these problems of the early arithmetic books are not motivated by an arcane interest in gambling but because they are permissible discussions of something that is formally equivalent to an impermissible enquiry. Money, the church held, cannot grow, so it is improper even to compute fair divisions of the growth of money, for that would entice Christians into speculation and usury. Gambling in contrast was merely frowned on, as an activity conducive to bad morals and in itself a mild sin. So the old textbooks of reckoning for merchants discussed the division problem as a disguise for what is really a discussion about merchant banking. That explains for example why such texts for the trading classes should have these computations about games.

I do not know whether Schneider is right, but I welcome the idea. Clearly if we should find the same division problem cropping up in Arab texts which were (as is commonly the case) the precursors of the early Italian ones, then this explanation will not hold up. So the suggestion is a testable one, although it will be some time before we can test it, for we do not now have a good understanding in full detail of the transmission of arithmetical lore from North Africa to Italy.

Schneider and I are quite agreed that the division problem, in early arithmetical texts for trade, does not indicate a pre–1650 "probability calculus." (Here he is disagreeing with van der Waerden.) The computations, he claims, had little to do with evidence of frequencies or other probabalistic ideas, and they occur at all only as a disguise for some problems about trade. Although his immediate motives have been different, Schneider thus furnishes support for my claim, that the early probability computations don't establish an early idea of probability. We connect them with probability only by hindsight. (Schneider recalls that in *Emergence* I claim that the word "probability" is associated with such computations only in the Port Royal Logic of 1662.) Had I known of Schneider's idea five years ago, I should have gladly incorporated it in my own analysis.

Among the other topics developed by Schneider I shall mention only Leibniz. It is natural for an historian of mathematics such as Schneider to investigate lines of filiation: who thought of what and passed it on to whom? My ample discussions of Leibniz must suggest to a reader with those interests that I think of Leibniz as having some effect on the history of probability. I had no such intention. I used Leibniz as what I called a "witness" to the new probability. He was indeed the first philosopher of probability. But I was using Leibniz in a somewhat subversive way. One of my theses was that the emergence of probability is largely an anonymous event. I was much impressed by the way in which probability ideas emerged in Holland, France, England and the German principalities, flowing from many pens, simultaneously and almost spontaneously. Necessarily one recites names of authors of texts in recording these events, but I hoped to suggest that these were merely representatives of a wider sphere of discourse. Leibniz serves me perfectly, not because he was a significant contributor to probability, but because he was not. He utters lots of the probability sentences, perhaps slightly ahead of his time, because that is what is being said, or what will be said. His connection between probability and the law occurred when he was isolated in an obscure principality, when he was adolescent. At most he got the idea from Weigel in Jena. Much later he corresponded with everyone and so became a convenient mirror of the new probability style as it was evolving. His own thoughts were all programmatic. His serious attempts at detail are full of mistakes. He had, I think, no serious influence on anyone. Schneider rightly takes me to task for implying that he had some influence on Jakob Bernoulli in this connection. Schneider rightly insists that he had none. So much the better for my use of Leibniz as a mere witness.

2. FAGOT'S PROBABILITIES AND CAUSES

The topics discussed in Fagot's paper begin a century after Schneider's end, and she takes us up to the 1871 siege of Paris. At first glance it may seem a long way from *The Emergence of Probability*, but for two reasons I am particularly glad that it forms part of this symposium. First of all, her essay is one chapter of her many-faceted enquiry that will illuminate the medical conceptualization of death and its causes. It is of fundamental interest to general philosophy, for it is directed at one line in the formation or, transformation of our idea of cause itself. It is not "archaeology" in the manner of Michel Foucault but it has learned from it, just as I have. So I am glad to have this part of Fagot's dissertation here, for it resembles my own methodology more than anything else we have seen this week. I welcome it for a second reason: it examines one aspect of the idea of probability in the very period about which I am now thinking and reading.

In the course of the nineteenth century there occurs the single most important transition in the idea of probability after 1660. Chance, said Hume (and everyone else) is a matter of hidden and secret causes. In the nineteenth century, statistical laws finally become what I call "autonomous," able to stand on their own without any necessity of postulating an underlying world of deterministic causes. One notable sign of this is that statistical laws are finally used not only to predict phenomena but also to explain them. (Thus Francis Galton in the 1870's is explaining facts about heredity such as "regression towards mediocrity" by the derivation of the correlation and regression phenomena from the assumption of a Gaussian or "normal" distribution of character traits.) Moreover, whereas at the beginning of the nineteenth century no one would disagree with Laplace's famous enunciation of absolute determinism, by the end of the century C. S. Peirce is only mildly radical when he claims to find no evidence or reason whatsoever for the "doctrine of necessity." Laplace lived in a world of deterministic cause but Peirce lived in a universe of chance. Most people imagine that this transition has something to do with quantum mechanics, but reading of the nineteenth century shows everything was set in motion long before.

It is a paradoxical time. Pedersen reminded us yesterday that W. K. Clifford announced that physics has no room for causes. Today Fagot tells us that at the very same time the medical instuitions produce their monstrous instructions for completing death certificates, adumbrating longer and longer lists of causes of death. These causes are to be exhaustive; one can die only of these causes and not even of "old age" which does not occur on the forms. Fagot argues that these causes prefigure a popular contemporary philosophical view that a cause of X is what increases the probability of X. That is, they are part of the *erosion of determinism* that occurs in the nineteenth century. Fagot has written part of the story of that transformation.

In this discussion I should restrict my comments to ways in which Fagot's work bears on the general thesis of *The Emergence of Probability*. I claimed – doubtless overstating the case, but thereby producing a testable thesis – that there was *no* probability before about 1660, and that after 1660 all the structure of probability concepts was fixed, with one exception. This exception is that the "erosion of determinism" and the associated "autonomy of statistical law" become established only as the nineteenth century develops. This is, I think not only consistent with Fagot's story but also illustrated by it. Consider her discussion of Laplace and the mortality curve.

The *Theorie analytique* uses two modes of inference that we take to be totally distinct, one Bayesian, and one "Bernoullian" (with a clear implication of a sense of confidence estimates in the manner of Neyman). The close reader is faced by a seeming paradox, for Laplace does not even bother to work out whether these two modes of inference are consistent, although he would have seen that in the long run they do coincide. The reason is that he thinks of them as part of the same conceptual apparatus which is, at bottom, founded on the same phenomenon of deterministic causes acting on similar cases so as to produce a frequency distribution. When we are ignorant of the causes we use a distribution that represents, for us, how we think that the unknown causes are arranged, a distribution that is robust, i.e. gives the same results for large numbers of different underlying distributions. Inverse inference from effects to causes is, for Laplace, always an attempt to discriminate equal from unequal causes. In his time everyone assumed that these causes are deterministic, acting of a universal necessity. Laplace's own favourite success story is the problem of why the proportion of newborn females registered in Paris is different from the rest of the country. Statistical analysis shows that the effect is significant: a hidden and secret cause must be in action. Essentially nonstatistical enquiry then shows that large numbers of peasants have been dumping their daughters at the city orphanages, keeping their sons to till the land at home. Statistics has pointed to a deterministic cause.

It is very difficult to transfer this easy success to the study of mortality. It is apparent that mortality rates differ, change, and can be affected. Yet nowhere can we lay our hands on a purely deterministic cause. There is regular reference to urn models from the time of Jakob Bernoulli and Leibniz, and many citations in the post-Laplacean period are given by Fagot. But even if we model the vicissitudes of life on an urn, how are we to conceive of changing urns? If each person has his own urn, as it were, what are the laws that determine that housepainters have one kind of urn, poets another? What changes in a person's urn when he inherits a fortune and quits painting houses for a living? Fagot cites Laplace groping after some master-urn, as it were, from which all the particular urns of life and of lives are drawn. We wonder at this incoherent speculation. It is explained by the necessity, for Laplace, of always postulating a hidden and secret cause. That cause must, of itself, be a deterministic one, or an intersection of deterministic causes.

All this changes as it becomes possible to speak of statistical laws in an autonomous way. A space is cleared for causes that act only probabalistically. I believe Fagot is correct when she concludes that, by 1871, the remarkable panoply of "causes" that appear on the medical forms are probabalistic causes. That is, they are causes in the sense that they substantially increase the likelihood of death, not that they produce death by the action of some universal and unfailing law.

I have already remarked that these are paradoxical times. It might seem as if the making of chances into real autonomous parts of nature, rather than mere epiphenomena, should greatly increase the sphere of freedom and diminish that of law. Only a very sophisticated audience could turn the whole thing on its head, and say, with some students of quantum mechanics, that even in microphysics we have determinism – the determinism of the ψ-function itself. Yet this move was made from the first, in the middle of the nineteenth century. As soon as Quetelet could produce, as statistical law, a function from which he could predict the number of suicides in Paris each year, then the spectre of free will was enjoined. The soul's decision to choose death is the ultimate act of freedom, the ultimate crime, the ultimate responsibility. Yet here is Quetelet showing that it is determined in advance almost exactly how many people will kill themselves in Paris next year! The sphere of determinism seems to expand rather than to contract. I would use this as an example of how the new philosophical problems, arising after a concept has been transformed, appear right from the start, and merely recur from time to time as new discoveries make one think of the same problem again.

I believe that if Fagot were to examine the more technical statistical literature she would find ample confirmation for her research. The various laws of large numbers provide a case in point. Bernoulli had shown that with a fixed chance p of "success" the proportion of successes in n trials tends to be about p, with the error diminishing as n increases. Poisson had asserted that one will obtain a limit even if p itself is chosen at random. This was proven by Chebysheff in 1846. In fact, the rate of convergence is actually quicker when p itself is drawn according to some uniform chance distribution. Bienaymé had investigated what he called the duration of causes. This is the case in which one has a sequence of k trials at chance p, followed by k more at chance p^*, and so on, with the chances themselves being drawn at random. In this case the rate of convergence is slower than with identical and independently distributed random variables. Bienaymé was trying to provide some analysis of the mortality curve when the curve changes; he thinks of various causes "enduring" for a period of time and then being replaced by others that endure, and he makes the simplest model of this. Bienaymé (who was in general loyal to the patterns set by Laplace) is, I think, still working along the Laplacian lines. He is not yet ready to countenance probabilities of death which are not organized by underlying deterministic causes. This is after Cournot has been groping along in that direction (I do not think that Cournot got as far as Fagot is inclined to suggest).

One of the most familiar facts about nineteenth century probability theory is the repetition of discoveries, and the problem of misattributed discoveries. The so-called Bienaymé-Chebyscheff inequality is a case in point, having been proven by the French author fifteen years before it was stated in full generality by the Russian one. Even the law of large numbers was ill understood; it was obscure even whether it was a mathematical or empirical fact. I believe that most of the misattributions arise not from malice or chauvinism but from the fact that no one was able properly to conceptualize what had been discovered. Even when the formalism could be written down with some rigour, no one knew what it meant. It was always supposed ultimately to be intelligible in terms of underlying deterministic causes. Only when that demand had been abandoned would one understand modern probability theory. Fagot's enquiries provide a detailed elucidation of one aspect of that transformation in the concept of probability. It is connected with perhaps the most fundamental philosophical change of modern times: the event that I call the erosion of determinism.

3. KOSTIOUK ON INDUCTION

The seventeenth and final chapter of *Emergence* is about the sceptical problem of induction, a problem which I attributed to Hume, and which is not discussed in earlier chapters dealing with earlier times. Kostiouk thinks this is a mistake. If we are to associate a philosopher with this problem, he suggests, it had better be Locke than Hume. Kostiouk carefully notes the passages from Berkeley which I quoted, and repeats my quotation from Foucault, that, in the light of these passages, "Hume has become possible." He says that Berkeley has nothing to do with the transition to a problem of induction, and cites three conditions, satisfied earlier than Berkeley or Hume, which suffice for a problem.

I think there may be less disagreement between Kostiouk and myself than meets the eye. At the beginning of the chapter I distinguished an "analytic" from a "sceptical" problem of induction. The former is the problem of trying to analyse inductive inferences, and discover the canons of the good and the bad. Having coined the phrase, "Leibniz-Carnap programme for inductive logic," I evidently think that *this* problem arose before Hume. Perhaps it would not have seemed a pressing or even a possible problem before Kostiouk's three conditions are met, but they were certainly met in the seventeenth century, and we do find an analytic problem then and maybe earlier. As he says, not even probability has much to do with the possibility of such a problem.

The analytic problem of induction is arguably the important one; the sceptical one, with its strong idealist overtones, has doubtless had too much of an interest for some Western philosophers. I do not wish to comment on its importance, but only to explain its existence. I believe that Kostiouk and I do not disagree about the facts, for although he does use my phrase, "sceptical problem," what he says about it is what I say about what I call the analytic problem.

Moreover, I too say that probability is not the essential ingredient in Hume's sceptical problem. Probability is there in 1660, but Hume publishes only in 1739. Something else must be invoked. I say that the fundamental transition in the concept of cause that is wrecked by the scientific revolution creates Hume's problem. In the 1660's we can find Glanville recalling many of the claims of the ancient sceptics, and posing a need for an analytic problem of induction. But it does not occur to him for a moment that there are not necessary connections in nature. That possibility is made manifest only by the Berkeley-Hume axis. Berkeley says that the so-called *causes* investigated in natural science are not causes but only *signs* that *Y* will follow *X*. There are indeed still causes, but they are the workings of God, not items uncovered by the scientist. Hume makes one further twist: yes indeed, the causes investigated by the student of nature *are* signs. Yet tradition is still right: the student of nature studies causes. All causes are only signs! There is no necessary connection! There is a sceptical problem of induction not because (as with Glanville) we may be in doubt as to whether we have located the necessary connections that will guide our predictions about the future, but because we now think there are no necessary connections, not even unknown ones.

In many ways Hume is closer to our time than many of his successors. He does assert the mainstay of his time, that chance is only a matter of hidden and secret cause. But he also considers the problem of "contrary causes," and suppose that many chance phenomena are the product of the war of contrary causes, each, we might suppose, with a certain power or strength. Thus there is more of a place in Hume for autonomous chance than there is in any other text for over a century. When Hume first published he was enthusiastic about his thoughts on probability, but in later writings he drops all that. I have suggested elsewhere that Hume is full of missed opportunities about probability – or rather, his various forays are undertaken before there is the conceptual space that will make sense of them[4].

4. NOVÝ ON THE CALCULUS OF PROBABILITY

Nový is right to observe that "the study of the probability calculus in the 18th century developed mostly in close association with heterogenous problems, for the accurate and mathematical solution of which it created a common base." It is not clear how this contrasts with probability mathematics at any other time, unless we contrast that branch of measure theory, advanced by Caratheodory and Kolmogorov, with applied probability mathematics. After Kolmogorov there plainly is a well defined and relatively independent branch of mathematics, and perhaps this is the contrast that Nový has in mind. At any rate the account of the *Encyclopédie* admirably confirms Nový's summary, although one would want to augment it, and to give the full breadth of the "heterogenous disciplines" that feed into these matters, such as D. Bernoulli's anticipations of statistical mechanics, or Mitchell's use of probability for analysis of astronomical data.

Nový concludes with a "marginal" remark on Bolzano that I find very interesting. I had not noticed the passage of Bolzano that he cites. It fits very nicely into that stream of development that I called the Leibniz-Carnap programme, and the lines of filiation are clear: Bolzano was very much influenced by the Leibniz-Wolffian school of philosophy that persisted even past Kantian and Hegelian times. And the account of probability quoted by Nový appears just to be the measure function that Carnap calls c†, and which Carnap attributed to Wittgenstein's *Tractatus*.

5. COHEN'S BACONIAN PROBABILITIES

Cohen has, like Fagot and myself, philosophical goals that extend beyond the recounting of historical events. His recent book offers a new theory about some kinds of inductive inference, and he wishes to remind us of the antecedents for his ideas. Two parts of his doctrine matter to us here. There is the *variation-of-circumstances* thesis, that a generalization gains empirical support from the independent variety of trials to which it is subjected. Then there is the *nonadditive probability* thesis, that there is a nonadditive concept of numerical probability which is the natural measure of the support alleged by the preceding thesis. Of course there are many other aspects of Cohen's philosophy, such as his claim that legal process requires a nonadditive probability concept, but despite his mention of a nineteenth century Scots advocate, Cohen does not at present seek earlier antecedents for that part of his theory.

There is no question that the variation-of-circumstances thesis has a distinguished pedigree, which Cohen traces at least back to Francis Bacon. What is in question is whether that has anything to do with probability. Cohen justifiably finds some of my own brief observations about Bacon a little cavalier. I said that "it is plain that Bacon had little interest in Humian induction and no concern with probability." Cohen thinks that Bacon did have a concern for probability, and that Cohen's nonadditive probability is the best explication of Bacon's conception. I shall address myself to three questions. (1) Is Cohen right about Bacon? (2) Is Cohen successful in finding predecessors for his theory of nonadditive probability? (3) How does Cohen's theory bear on the theses of *The Emergence of Probability*?

My answers to these questions, to be developed below, are as follows. (1) Cohen does usefully direct our attention to some aspects of Bacon that are too frequently ignored. But although I was wrong to say that Bacon had "no concern with probability," the contrast with the post-1660 era, including Bacon's own professed disciples, such as Hooke (cited by Cohen), is very great. We might correct my exaggerated claim by saying that Bacon had *less* concern with probability than *any* philosopher of science after 1660 – less even than Kant! (2) There is a real tradition of nonadditive probability completely ignored by Cohen. Cohen's attempt to find antecedents for *his* concept of nonadditive probability compares unfavourably with Glenn Shafer's historical research on a nonadditive concept. (3) Cohen's useful correctives about Bacon do not seem to bear on the theses of my book at all. I now proceed to some details.

Francis Bacon has had an uneven reputation. He is in effect the patron saint of the Royal Society, yet what is said about Bacon in the early days of the *Philosophical Transactions* has more to do with a certain experimental spirit attributed to Bacon. In those days he was perhaps too much praised for what he had never in fact said. Recently, in contrast, he has been somewhat condemned for an inductivism that is not his. We are grateful to Cohen for getting back to the texts.

Bacon, I wrote, "does not aim at inference under uncertainty." Well, as Bacon himself said, if a man begin in certainty he shall end in doubt, and of course I have to retract what I said. In the natural sense of the words it is too strong; I had wrongly been using the words in that technical sense pre-empted by statisticians who write of "inference under uncertainty." In the natural sense of the words Bacon *was* concerned with inference under uncertainty: as Cohen quotes him, his method was to be one of "setting up degrees of certainty." This is what Cohen calls a "programmatic" announcement from the Preface to the *Novum Organum*. Unfortunately it does require "interpretation" (to use Cohen's word) for Bacon does not set up anything that he goes on to call degrees of certainty. It is instructive to notice that after 1660 writers hark back (anachronistically, I claim, against Schneider) to Carneades or the New Academy. Bacon, in contrast has little use for the grades of certainty or of "probability" of those writers (e.g. *Aphorism* LXVII), but he does not replace them by anything of his own. When he does reach what appears to be certainty he contrasts it with probable reasoning (e.g. in Book II, §xxxvi, after discussing the *Instantiae crucis*, he contrasts such definite conclusions with those reached *per rationes probabiles*).

Bacon never delivers the "degrees of certainty" announced in his preface. Cohen has a plausible conjecture about what Bacon might have meant. The degrees of certainty must be construed in terms of the ladder of axioms that pass from the less general to the more general. Cohen connects this in an attractive way with Whewell's theory of the consilience of inductions, according to which a proposition supported by several different kinds of evidence ranks higher than one supported by fewer kinds of evidence. It is not sheer quantity of evidence but variety that counts. Cohen argues that this idea cannot be cast into the mould of additive probability, but requires a nonadditive framework. Hence it was wrong of me to say, after quoting Bacon, "there is little room in this conceptual scheme for a working concept of probability." After all Cohen has just shown us a working concept placed within his version of Bacon's scheme. I continue to believe that there was no place for it in the scheme as understood by Bacon and his contemporaries; at any rate Cohen and I agree that before 1660 no one elaborated any quantitative concept of probability. Certainly Bacon did not.

Cohen urges that we should not think of additive numerical probabilities as the only kind of probability. It does not follow from the fact that Bacon made no use of these, that he had no probability ideas. I agree. I myself deplore the hegemony of one kind of probability, and believe, for example, that, in the domain of frequency-style reasoning, the technically defined ideas such as degrees of significance, likelihoods, and probability are all important measures of support for hypotheses. There is no sense to adding likelihoods, and only fortuitously can one sometimes combine degrees of significance to get meaningful numbers. It is not a fixation on any particular probability concept that prevents me from finding probability in Bacon. I do not think it is there, and am not convinced by Cohen's quotations. I do grant that there are Baconians with a probability concept. Cohen cites Hooke and Boyle. But so much is just one of the theses of *Emergence*: after about 1660 you find probability on every pen, while before you find it on none.

Now I turn to my second question. Does Cohen succeed in finding historical antecedents for nonadditive probability? Cohen takes Butler to provide the first instance of a clearly nonadditive concept. In fact nonadditive probabilities are very clearly enunciated and investigated by Jakob Bernoulli a generation or two before. Cohen calls additive probabilities Pascalian. I know of no evidence that Pascal was committed to the view that all probabilities are additive. Glenn Shafer, who has worked more intensively on this question, thinks that Bayes is the first to insist that *all* probabilities are additive, and so he calls "Bayesian" what Cohen calls "Pascalian."[5] Let us speak simply of additive and nonadditive probabilities.

Cohen has one idea of nonadditive probability, Terence Fine[6] another, and Glenn Shafer a third. In addition anyone concerned with probabilities on an interval scale is likely to have at least a passing interest in nonadditive probabilities, Shafer is the proponent of nonadditive probabilities who has provided the most elaborate historical research. He points to Bernoulli's use of them, and to H. G. Lambert's rejection of Bernoulli's analysis. These matters were discussed in the nineteenth century, for example by Todhunter and even De

Morgan. Shafer can with some justice speak of a submerged tradition of nonadditive probability.

Bernoulli gives an example in which the probability of a proposition *p*, plus that of its negation, may exceed unity. His case is not interesting because it looks as if one could normalise to obtain an ordinary additive probability, but another class of cases is more instructive. He carefully distinguishes two kinds of evidence, which he calls "pure" and "mixed." Mixed evidence is the familiar kind of which is suggested by frequencies and partitions. It arises, say, when 3/4 of the cases in a problem support one conclusion, while 1/4 support its opposite – or when we know just that 3/4 of the outcomes from a trial fall one way, and 1/4 the other way. In such situations of mixed evidence (some evidence for a conclusion, some against it) we naturally use additive probabilities.

Pure evidence is of a sort that counts for a conclusion but says nothing about the opposite. Witnesses provide an example. *W* is a disinterested witness who usually gives well-founded testimony. When he does that, then if he says *p*, then *p* is true. But one time in five his testimony is ill-founded. On those occasions what he says is simply irrelevant to the truth of *p*. (This is different from the witness with an interest in the proceedings, who actually lies, in a well-founded way, some of the time.) Suppose that *W* says *p*. In such a case of pure evidence, 4/5 of the total evidence supports *p*, and 1/5 says nothing. We recall Bacon speaking of degrees of certainty. Bernoulli does something with that idea. He says that we have 4/5 of certainty for *p* – from which it does not follow that we have 1/5 of certainty for not-*p*. And Bernoulli calls such a "degree of certainty" or "part of certainty" a probability. So Bernoulli thought some probabilities are nonadditive.

Bernoulli addresses the problem of how to combine additive and nonadditive probabilities. What happens when we have both pure and mixed evidence for a given proposition? Bernoulli arrived at one answer. H. G. Lambert thought that Bernoulli was wrong, and in effect reasons as if all probabilities are additive. Most subsequent writers have sided with Lambert. Shafer makes a strong case that Bernoulli was right. I have myself expressed some reservations about Shafer's solution.[7] But any rate here we have a tradition. Moreover Shafer, by drawing our attention to the Bernoulli-Lambert debate, has used the history to teach us something about nonadditive probability. Here, he finds, is a class of problems, clearly posed, in which different theories of probability give different answers.

Cohen's attempt to find historical antecedents gives us nothing comparable to Shafer's work. He ascribes an interest in nonadditive probability to Butler, some 21 years after the posthumous publication of Bernoulli's *Ars conjectandi.* Cohen quotes Butler supposing that there would be a strong presumption for an hypothesis *p*, yet none for not-*p*. Indeed what Butler has in mind is exactly what Bernoulli had called the pure situation, rather than the case of mixed evidence. Like so many of us, Butler had his fish to fry. He thinks that there is no presumption against the hypothesis, "that there is an intelligent Author of Nature, and natural Governor of the world." He thinks that there are many occasions on which that hypothesis has been "proved with accumulated evidence" (analogy, abstract reasonings, tradition, testimony and general consent). This suffices, he thinks for us to act on the belief in God. Moreover Butler guards himself against those who say that the proofs of God are inadequate. Even if they furnished but little probability for God, there is no presumption at all that God does not exist. So Butler supposes that there can be zero probability for God's non-existence, and perhaps only small positive probability for God's existence. At least, that is what follows if we substitute numbers for Butler's words.

Cohen does not refer to Butler in his book, but the quotation is his paper might serve as some basis for speaking of "Pascalian" probability. For Butler follows the innovation of Pascal in ascribing probabilities to the hypothesis of a natural Governor of the world. Even if Butler had not read the end of Port Royal *Logic* where Pascal's wager is sketched, he was familiar with the coarser form of the argument in Locke's *Essay* II. xxi. 72. Pascal assigns betting rates to the existence of God, and uses the additivity of those rates to argue from expectation. There are in fact three arguments in Pascal, a fact glossed over by Locke. Now Butler is certainly doing something non-Pascalian, and perhaps he does so just because of criticism directed at Locke's *Essay.* Pascal's numbers are additive, being derived from betting rates; if we assign numbers to Butler's probabilities and presumptions, they will be nonadditive. If we must invoke Pascal I think we should speak not of Pascalian and non-Pascalian probabilities, but of Pascalian and non-Pascalian probabalistic arguments for the existence of God.

Cohen is, then, right to find a nonadditive probability in Butler, but fails to mention Bernoulli's more explicit assignment of probabilities that do not add to one. If, like Cohen and Shafer, one seeks historical antecedents for nonadditive probabilities, one must locate the origin in Bernoulli. But unfortunately when one follows out the details of Bernoulli's conception, one seems more led to Shafer's present theory than to Cohen's. That, of course, shows nothing about the relative merits of the modern theories.

I conclude with my third question. Cohen began saying he would correct a point in *Emergence.* Does the correction bear on any of the theses of that book? I don't think so. At the beginning of this set of replies I describe the archaeological enterprise that governs the book. It is almost deliberately the opposite of Cohen's paper. He is seeking a precursor, whereas, as I wrote, my model made me "tend to disregard the anticipation, the man who, with subtle interpretation, can be presented as a precursor of the modern way of thought." With subtle interpretation Francis Bacon can be made to anticipate a certain idea of probability. It requires no interpretation to find probability in the Baconians whom Cohen quotes after 1660. This contrast is an example of precisely the phenomenon that motivated *Emergence.*

REFERENCES

* The papers by Fagot, Schneider, Nový, and Kostiouk appear above in this volume. As indicated in the Preface, L. J. Cohen's paper has appeared in the *Journal of the History of Ideas.* (Editors' note.)

[1] Miles Burnyeat, 'Carneades was no probabalist,' to appear in a projected volume, *Riverside Studies in Ancient Skepticism*, ed. David K. Glidden.

[2] Roderick Chisholm, 'Sextus Empiricus and modern empiricism,' *Philosophy of Science* **8** (1941), 378.

[3] Daniel Garber and Sandy Zabell, 'On the emergence of probability,' Technical Report no. 76, Department of Statistics, University of Chicago, to appear in *Archive for the History of Exact Sciences.*

[4] Ian Hacking, 'Hume's species of probability,' *Philosophical Studies* **33** (1978), 21–37.

[5] Glenn Shafer, 'Non-additive probabilities in the work of Bernoulli and Lambert' *Archive for the History of Exact Sciences* **19** (1978), 309–370. *A Mathematical Theory of Evidence*, Princeton, 1976.

[6] Terence Fine, *Theories of Probability*, New York, 1973.

[7] Ian Hacking, 'Combined evidence,' in S. Stenlund (ed), *Logical Theory and Semantic Analysis*, Dordrecht 1974, pp. 21–37.

WESLEY C. SALMON

JOHN VENN'S *LOGIC OF CHANCE*

The present paper is offered largely in the spirit of an historical conjecture. It is my view that Venn holds a place of great significance in the history of the philosophy of probability, and that the importance of his contribution is not generally appreciated. The main aim of this paper is to outline this contribution and argue for its significance. I have no serious doubts as to the importance of the ideas to be found in Venn's work, but I cannot say with complete confidence to what degree they were unanticipated by other authors. To the extent that I am familiar with the work of his predecessors, it strikes me that Venn's contribution was highly original, but I readily concede that my knowledge in this respect is far from exhaustive. Let me, therefore, set out what I take to be the most significant aspects of his work in the hope that others who are more knowledgeable in the historical aspect will be able to correct any historical injustices I may commit.

Venn is best known today for his diagrammatic technique which makes it possible for any child with a stick and a sandbox to test the validity of any standard categorical syllogism. His contributions to the philosophy of probability are, in my opinion, far more significant. During the period 1866–1889, he published three books on logic (at any rate, each had the word "logic" in its title). The first of these, *The Logic of Chance* (first edition, 1866), offers a detailed, sustained, and systematic analysis of the concept of probability. To the best of my knowledge, it is the first work in which the frequency interpretation is advocated and analyzed at such length and in such detail. Given the pervasive influence of the frequency interpretation on subsequent work on statistics, probability, and induction, it seems to me that this book constitutes an epoch-making work. It is, I suspect, insufficiently known or appreciated. The second book, *Symbolic Logic* (1881), is a standard sort of text on deductive logic. It seems to have no particular significance beyond the diagrams (see Kneale & Kneale, pp. 420–422). The third work, *The Principles of Empirical or Inductive Logic* (first edition, 1889), treats inductive logic after the manner of

J. Hintikka, D. Gruender, and E. Agazzi (eds.), Pisa Conference Proceedings, Vol. II, 125–138.

John Stuart Mill. In my Ph.D. dissertation (UCLA, 1950), I analyzed Venn's work on probability and induction at some length, and concluded that, partly because of certain basic misconceptions regarding the nature of logic and partly because of excessive devotion to the work of Mill, Venn's contributions to inductive logic are not especially significant.

All of Venn's "logic books" apparently became standard texts during the latter part of the nineteenth century. *The Logic of Chance* went through three editions – 1866, 1876, 1888. The "fourth edition" (Chelsea, 1962) is merely a reprinting of the third. However, the best known frequentists of the twentieth century, von Mises and Reichenbach, whose earliest works on probability date from the teens of this century, give Venn only passing mention; they did not appear to be genuinely influenced by him. When I undertook to write my dissertation on Venn under Reichenbach's supervision, I had the impression that Reichenbach had no first-hand knowledge of Venn's work on probability. Keynes (1921) takes Venn's views quite seriously, and devotes a substantial amount of space to discussion of them, but he is, of course, intent upon rejecting them. I do not know whether Keynes's critique was historically responsible for the subsequent neglect of Venn, but I am not aware of any other important author giving him much consideration thereafter. Venn's critiques of the subjective and classical interpretations of probability have been included in a couple of recent anthologies (Kyburg & Smokler, 1964; Madden 1960).

Venn obviously did not invent the frequency interpretation of probability. The basic idea had been expressed by Ellis, Cournot, and, indeed, even Aristotle. But in the light of Venn's searching and detailed elaboration, it seems to me that earlier attempts at articulation should be considered anticipations rather than clear formulations. As a matter of fact, it seems to me that most authors prior to Venn had little appreciation for the problems of interpretation of the probability concept. Their statements were usually brief, enigmatic, and equivocal. To the best of my knowledge, Venn was one of the first authors to take the problem of interpretation seriously, and to attempt to offer a carefully reasoned systematic solution – at the very least, I believe, he was the first frequentist to do so.

There is, of course, something anachronistic in talking about "interpretations" of probability in the nineteenth century, if we construe that term in connection with axiomatized versions of the calculus of probabilities. These did not exist until the twentieth century. Making use of this latter-day notion, however, we can introduce a distinction which is central to the thesis of the present paper. We may contrast the "frequency interpretation" of probability with the "frequency theory" of probability. Those who adhere to the frequency *theory* I shall call "frequentists." In modern terms, we may note that Carnap recognizes the frequency interpretation as one of two admissible interpretations of the probability calculus. Carnap is not a frequentist. The frequency *theory* of probability is the thesis that *all* legitimate uses of the probability concept can be reduced to, or analyzed in terms of, frequencies. The claim that either finite frequencies or limits of frequencies in infinite sequences constitute admissible interpretations of the probability calculus is widely accepted at present – taking a few niceties into account – and I shall not discuss it any further. The frequency theory, as just characterized, is by no means widely accepted.

It seems to me that a distinction of the foregoing sort can reasonably be read back into the nineteenth century in the sense that many nineteenth century authors acknowledged, explicitly or implicitly, that probability sometimes had something to do with long run frequencies. How could it be otherwise, when the instigation for the study of the mathematical theory of probability had come out of the long sequences of trials in games of chance? There is, of course, an important difference between vague formulations of the frequency interpretation, such as Aristotle's "The probable is that which happens often," and precise formulations, such as "Probability is the limit of the relative frequency in an infinite sequence of events." Venn provided a clear formulation of the frequency interpretation, but I am not sure he was the first to do so. Even if he was, I do not consider that his most significant achievement. What was even more important was his adoption and defense of the frequency *theory* of probability.

Hacking (1975) has made a strong case for the historical thesis that the concept of probability could only emerge in the presence of two sets of ideas – those relating to relative frequencies and those relating to degree of evidential support. These two kinds of notions are directly reflected in Carnap's probability$_1$ (degree of confirmation) and probability$_2$ (relative frequency). Venn, of course, was interested in relative frequencies; Keynes takes him to task for slighting degree of confirmation. One can, however, endorse this view – insisting that the concept of probability has important applicability in the context of degree of confirmation or degree of reasonable belief – without giving up the frequency theory. The frequentist can adopt this posture if he is willing to argue that all of the various legitimate uses of the probability concept can be analyzed in terms of frequencies.

Von Mises is not a frequentist in this sense. He holds that the only legitimate meaning of the probability concept is given in terms of the limit of the relative frequency in an infinite sequence of events, but instead of analyzing other uses of the probability concept in frequency terms, he simply condemns them as incorrect. In particular, he maintains that it makes no sense to speak of probabilities of single events or probabilities of scientific hypotheses. This view is, at best, a Procrustean version of the frequency theory. Reichenbach, in contrast, has not denied the applicability of probability concepts in connection with the single case or the support of hypotheses; he maintains, instead, that the frequency interpretation can supply a concept adequate to these contexts. One might complain that he was not successful in showing that the frequency interpretation has such a capacity, but there are no grounds for complaining that he did not try. Venn shares the same general viewpoint as Reichenbach – at least to a degree. He succeeds rather well, I think, in treating the problem of the single case; concerning the problem of the probabilities of hypotheses he is less successful. But even here, it seems to me, he manages to bring out some fundamental issues. Venn can thus be viewed, I think, as the first systematic exponent of the frequency *theory* of probability. Whether one accepts or rejects the frequency theory, it seems to me that Venn's careful statement and systematic defense of this view has great historical significance.

Venn sets out the fundamental features of his theory of probability in a straightforward fashion. Probability deals with series of occurrences which display individual irregularity but regularity in the aggregate. One constructs abstract mathematical sequences which have limiting frequencies, and establishes an approximate correspondence between these abstract entities and the large or potentially infinite sequences of actual concrete events. The basic mathematical rules of probability, such as the addition and multiplication rules, provide the means for performing arithmetical operations upon such limits. Venn is completely explicit in acknowledging the process of mathematical idealization involved in this procedure, and he clearly recognizes the factual emptiness of inferences performed by means of these arithmetical operations.

When we want to ascertain the probability associated with any particular concrete sequence, we do so by empirical examination of a portion of the sequence and by inductive extrapolation to the remainder. In his discussion of this problem Venn takes pains to argue against a priori and subjectivist approaches. He concludes,

> Experience is our sole guide. If we want to discover what is in reality a series of *things*, not a series of our own conceptions, we must appeal to the things themselves to obtain it, for we shall not find any help elsewhere. We cannot tell how many persons will be born or die in a year, or how many houses will be burned or ships wrecked, without actually counting them. In saying that experience is the sole guide, I mean of course experience supplemented by the aids that Inductive Logic can afford. (1866, p. 26)

Venn makes a strong case for the applicability of his frequency conception to situations – such as games of chance, insurance, and social statistics – in which large aggregates of events are involved.

Many authors would agree in large measure with Venn's characterization of the frequency interpretation of probability and its applicability in situations of the foregoing kinds. A frequentist (provided he wishes to avoid the heroic denials of von Mises about the applicability of probability in others contexts) must show how other uses of probability concepts can be explicated in terms of limiting frequencies. The most immediate problem concerns probabilities of single events. Venn gives what I consider in all essentials an adequate and appropriate treatment of this problem. In chapter III, section 33, of the 1866 edition, he offers an explicit definition of the phrase "probability of an event" and its synonym "chance of any given event happening" in terms of a frequency in a sequence. The topic of this particular chapter is "Gradations of Belief," and in it Venn distinguishes various psychological factors which may play a role in determining degrees of belief from the logical factors which can provide a basis for a *justified* degree of belief (cf. 1866, p. 112). At the end of chapter III he remarks,

> The above investigation describes ... the amount of truth which appears to me to be contained in the assertion frequently made, that the fraction of probability represents also the fractional part of full certainty to which our belief of the individual event amounts. Any further analysis of the matter would seem to belong to Psychology rather than to Probability. (p. 111)

In this chapter Venn gives, I think, a rather good critical discussion of problems associated with the measurement of degrees of belief, of the construal of raw subjective degrees of belief as probabilities, and of the use of such a priori devices as the principle of indifference as a basis for establishing justified degrees of belief. I tend to agree with the apparent judgment of Kyburg and Smokler (1964) that Venn's arguments are the sort which must be answered by any adequate contemporary subjective theory. Writing in 1866, Venn obviously did not have the sophisticated approach to measurement of degree of belief or the technical concept of coherence which advocates of recent personalist approaches have available to them. I think it is clear, however, that Venn would agree with a number of twentieth century probability theorists that, even if coherence provides necessary conditions for rational degrees of belief, they are by no means sufficient.

In chapter VII of the 1866 edition, Venn returns to the problem of the single case and addresses a crucial issue which was not adequately treated in the earlier chapter, namely, the fact that a given single event belongs to many different series, and the value we assign as its probability may thereby differ, depending upon the series to which we refer it. (1866, p. 175.) His resolution is precisely the same as Reichenbach later advocated: refer the single case to the narrowest reference class for which reliable statistics are available, or better, to the broadest homogeneous reference class. The nature of statistical relevance relations, and their importance in this context, were clearly articulated by Venn. Thus, it seems to me, Venn provided, in the mid-nineteenth century, a very creditable version of the frequency interpretation of probability, and of its applicability both to aggregate phenomena and to single cases. This was a non-negligible achievement, I believe, at that stage in the history of the theory of probability.

Unfortunately, Venn did not produce an equally adequate treatment of the probability (or confirmation) of hypotheses. This aspect of Venn's theory is more interesting than its more successful aspects, for there is more to be said than that, in the middle of the nineteenth century, he did about as much as anyone has done subsequently. It will be worth our while to see why he did not succeed as well in dealing with this issue.

In discussing Venn's treatment of the probability of hypotheses, it will be helpful to distinguish general hypotheses from particular hypotheses, and deal with the two types separately. Where general hypotheses are concerned, whether they be universal generalizations or statistical generalizations, Venn argues that probability is inapplicable. General propositions are established by induction, and on that basis they can attain what amounts to certainty for all practical purposes. This thesis is not stated as clearly in the 1866 edition as it is in the third edition of 1888, which appeared in the year just preceding the publication of his inductive logic book, *The Principles of Empirical or Inductive Logic.* Clearly, he had given a great deal of thought to inductive logic – though I fear we cannot say that his thought was clear – in the twenty-two years between the appearance of the first and third editions of *The Logic of Chance.* I discussed his inductive logic in some detail in my Ph.D. dissertation, and do not want to say much about it here. His approach to induction was strongly influenced by Mill, and he sought to base the whole of inductive logic in a series of postulates which would guarantee its reliability. This approach seems to me to be altogether unsatisfactory. Venn simply did not have any appreciation for the problems of confirmation which have become such a large concern in the twentieth century. Moreover, he did not have the tools with which to deal with these issues.

I am rather strongly convinced that the key to the problem of probability (or confirmation) of hypotheses lies in one way or another in Bayes's theorem. Venn was rather critical of a number of the theorems of the probability calculus, such as the Bernoulli theorem and the Laplace rule of succession, which are a bit more complicated than the simple addition and multiplication formulas. Moreover, he denied that the distinction between direct and inverse probabilities has any genuine significance. He rejected the Bernoulli theorem and Laplace's rule of succession on the ground that each of them illegitimately mixes induction with probability. His basis for this opinion was his claim that probability manipulations, by themselves, are factually empty and do not extend our knowledge from the observed to the unobserved. Both the Bernoulli theorem and the rule of succession seem to do just that – to involve inference from observed cases to unobserved cases. Consider the rule of succession, which Venn states as follows: "To find the chance of the recurrence of an event already observed, divide the number of times the event has been observed, increased by one, by the same number increased by two." Venn devotes an entire chapter (chap. VI. 1866) to this "rule" and criticizes it harshly, adducing many cases in which it is false as just stated. As we know, the foregoing statement is incorrect, for the theorem holds only on the condition that the prior probabilities of the possible frequencies are equal. Venn is *not* unaware that, under these conditions, the theorem is valid (cf. chap. VIII, sec. 4, 1866 ed). His point is that it is often stated without imposing the appropriate conditions, and it is frequently applied in circumstances in which the conditions are not fulfilled. In most situations, the assumption of equal prior probabilities is an extremely artificial assumption, but that did not stop people like Laplace from using the rule of succession to ascertain the probability of the sun rising tomorrow morning. There was, in the nineteenth century, a strong tendency to use the principle of indifference rather indiscriminately – e.g., to assume that, in the absence of knowledge to the contrary, prior probabilities can be assumed to be equal.

Venn objects to the Bernoulli theorem on similar grounds. This theorem is sometimes applied to calculate the probability of getting twenty sixes in a row with a standard die, and to infer the frequency with which such a result will occur in a sufficiently extended series of tosses. Again, Venn recognizes clearly that, under certain assumptions about the probabilities of the various outcomes and the independence of the trials, the calculation produces correct results, but he is highly critical of the use of the theorem in situations in which the conditions are not known to be fulfilled. Again, the principle of indifference is to blame; nineteenth century theorists were ready to pronounce types of events independent if there was no evidence to the contrary. Venn's criticisms of these theorems were not a result of mathematical ignorance on his part; they resulted from his healthy antipathy to a priorism. In this respect he stood in good company squarely within the tradition of British Empiricism.

Venn has another ground for complaint against the Bernoulli theorem. As he formulates it – in a manner not unusual in the nineteenth century – it says, "in the long run all events will tend to occur with a frequency proportional to their objective probabilities." (1866, p. 35) Since Venn is convinced that the only kinds of probabilities are frequencies, he objects to the notion of "objective probability" as another manifestation of unwarranted a priorism. He did not see that even the frequentist can accept this mathematically valid theorem, interpreting it to say that the long run frequencies of certain combinations (finite frequencies) have certain values.

Venn does, however, become involved is basic misunderstandings of inverse probabilities, the topic of chapter VIII of the 1866 edition. This fact, I suspect, precluded his understanding and applying Bayes's theorem in any manner – including, of course, as a confirmation schema. He introduces his discussion with a quotation from De Morgan: "In the preceding chapter we have calculated the chances of an event, knowing the circumstances under which it is to happen or fail. We are now to place ourselves in an inverted position: we know the event, and ask what is the probability which results from the event in favour of any set of circumstances under which the same might have happened." Venn offers two examples of the contrast between direct and inverse probabilities, one correct and one mistaken. The first is (1866, p. 204):

(1) A ball is drawn from a bag containing nine black balls and one white; what is the chance of its being the white ball?

(2) A ball is drawn from a bag containing ten balls, and is found to be white; what is the chance of there having been but that one white ball in the bag?

Showing that the second question can be answered only if the prior probabilities of the different possible frequency distributions are available, and noting the arbitrary character of the assumption of equal priors, he proceeds to calculate the answer. His treatment of this example is completely correct.

Venn unfortunately sees the distinction between direct and inverse probabilities as based upon temporal considerations – an easy impression to form when one thinks of inverse probabilities as probabilities of causes, given that causes precede their effects. Since he has argued in another context (1866, chap. 5, secs. 5–17) that temporal relations are irrelevant in probability theory, he is led to reject the distinction. He offers the second example in support of his thesis (1866, p. 209):

(1) What is the chance that if A.B. dies he will die of typhus fever?

(2) A.B. is dead; what is the chance that he died of typhus?

He rightly observes that (1) and (2) both ask for the same probability, namely, the conditional probability, given death, that the cause was typhus. Whether we ask for this probability before or after the death occurs (or we know of it) is irrelevant. This example is, of course, defective. The natural way to handle such an example is to distinguish

(1′) What is the chance that if A.B. gets typhus he will die?
(2′) What is the chance that if A.B. dies he will die of typhus?

The answer to (1′) would usually be called the direct probability, and the answer to (2′) the inverse probability. Clearly these two probabilities are not identical. To compute the inverse probability from direct probabilities by Bayes's theorem, it is necessary to know, in addition to the probability that he will die if he contracts typhus, the probability that he will contract typhus and the probability that he will die from any cause whatever. We must admit that Venn simply blundered on this example, and was consequently altogether misled regarding inverse probabilities.

By the time of the third edition of *The Logic of Chance* appeared, Venn had apparently become aware of the inappropriateness of his second example. He replaces it by another (1888, chap. VII, sec. 15):

(1) What is the chance that a Londoner will die of bronchitis during a given year? (Answer: 1/1250)
(2) A Londoner is dead; what is the chance that he died of bronchitis? (Answer: 1/25)

In this later discussion, Venn still maintains that there is no fundamental distinction between direct and inverse probabilities – i.e., both must be based upon statistical data. However, as a practical matter, it often happens that the statistical data needed to establish the inverse probability are not easily available. In such cases, Venn argues, we should simply withhold judgment rather than introducing a priori or arbitrary assumptions about the prior probabilities. He concludes his earlier discussion with this observation: "But as it is a hard thing to be deprived of the right of forming a positive opinion merely because we happen to have no grounds on which to rest it, we are driven to assume for ourselves a fictitious experience ... which opens the way at once to confusion and dispute." (1866, p. 211)

If Venn is right, we can handle universal and statistical generalizations by induction (without recourse to probability), and we can handle justified degree of belief in single events by assigning them to suitable series (a procedure which falls within the purview of probability). These claims lead to the usual objection that not every single case lends itself to referral to any series – at least not to any series for which we have statistical information. Historical events are often cited as examples. What is the probability that Brutus stabbed Caesar, or that Jesus rose from the dead after crucifixion? (Venn was a clergyman before he returned to Cambridge in 1862 to take up a newly created lectureship in Moral Science, and he exhibited considerable interest in miracles, improbable occurrences, and testimony.) Some of his most valuable remarks on the probability of hypotheses are to be found in such contexts. Political events are also problematic; what is the probability that Jimmy Carter will be elected to a second term of the U.S. Presidency?

We might appropriately consider propositions regarding these and many other sorts of events as particular hypotheses. If the events in question can be referred to appropriate series, we can assign probabilities to them, thereby justifying corresponding degrees of belief. Otherwise, on Venn's theory, we are out of luck. If we were to object that people do, in fact, assign probabilities to hypotheses of this sort, or have degrees of belief concerning them, Venn would reply that such degrees of belief – though perhaps real – are unwarranted by evidence, and that it is improper to associate such degrees of belief with probabilities. In response to these views, many authors – Keynes and Carnap among them I should think – would say that Venn's conception of probability is simply too narrow to cover the important and legitimate uses of probability. Add to that the fact that many would also claim – and I would be one of these – that Venn's treatment of general hypotheses in terms of inductive near-certainty is untenable, and we have another area which cannot easily be handled by Venn's conception of probability.

Carnap has, as we all know, insisted that there are two concepts of probability, degree of confirmation and relative frequency. Hacking has shown that, historically, the probability concept has two such aspects. Under these circumstances, it is tempting to say that there are simply two different concepts, and that it is superfluous and futile to try to analyze one of them in terms of the other. We need degree of confirmation to handle the probabilities of hypotheses, general and particular, and perhaps all single case probabilities. Philosophical analysis and linguistic usage, it might be said, agree in support of this point of view. Venn was, I believe, sensitive to this very issue, and his sensitivity grew in successive editions of *The Logic of Chance.*

In chapter XI, secs. 10–11, of the third edition (1888), Venn summarizes his views on the probability of hypotheses. The fundamental issue surrounds his insistance that assigning a probability inescapably involves an element of justification, and that justification necessarily involves an appeal to statistics.

> The best way therefore of describing our position in Probability is as follows: – We are *entertaining a conception* of some event, past, present, or future.... In its present condition it would be incorrect to call it a fact, though we would willingly, if we could, convert it into such by making certain of it one way or the other. But so long as our conclusions are to be effected by considerations of Probability only, we cannot do this. The utmost we can do is *estimate* or *evaluate* it. The whole function of Probability is to give rules for so doing. (1962, p. 276, Venn's italics)

The rules must provide justification.

> ... the process of inference is that of ascertaining to what extent we are justified in adding this conception to the already received body of truth and fact. (1972, p. 276)

The justification must appeal to frequencies.

> ... we always *justify* our conclusions by appeal to facts, viz. to what happens in the long run. (1962, p. 277, Venn's italics) By means of reference to statistics ... we are enabled to say how much this conception is to be believed, that is *in what proportion out of the total number of cases we shall be right in so doing* [i.e., taking the hypothesis to express a fact]. (1962, p. 276, my italics)

Frequency of truth thus constitutes the ultimate standard for the attribution of probabilities to hypotheses. This thesis, in my opinion, constitutes the core of Venn's most significant contribution to the philosophy of probability. I am inclined to think that, in some sense, Venn must be right about this. But whether one agrees or disagrees, I do not see how one could fail to look upon the clear articulation of this version of the frequency theory of probability as a profoundly important development in the history of the philosophy of probability.

In recent years, it has not been fashionable to discuss issues of justification in any very explicit manner. Many authors – Carnap, Goodman, von Wright, to name a few – have suggested that we should turn our attention away from problems of justification, concentrating instead on the elaboration of systems of inductive logic, inductive probability, or degree of confirmation. Venn, it seems to me, advances a pertinent challenge to this view. If degree of probability in the sense of degree of confirmation is to mean anything, it must signify the degree to which the evidence offers *rational* support for the hypothesis – the degree to which the evidence warrants or justifies acceptance (or something like that) of the hypothesis. Venn offered incisive arguments in principle against a priori assessments of probabilities. He did not merely argue that unbridled use of the principle of indifference leads to inconsistency; his arguments are philosophically telling against the most carefully regimented Carnapian use of that principle in assigning the initial measure function. He did not criticize Carnap for assigning weights in a way which leaves general hypotheses with zero degree of confirmation – of course, how could he have guessed such a thing would happen? – but his argument cuts equally against more sophisticated a priori assignments of probability measures such as those developed by Hintikka and associates. In the end, Carnap rested his final appeal upon inductive intuition. Venn said, as noted above, if we want to deal with the real world, and not just with our conceptions, then we have to tie probabilities with the real world – not merely treating them as articulations of our intuitions.

It has seemed to me that a good deal of the work which has been done on inductive logic and degree of confirmation has been inspired by the supposition that there is a relation of *partial entailment* underlying inductive logic in a manner quite parallel to the way in which the relation of *full entailment* underlies deductive logic. Although such an idea may have some heuristic value, I believe it is more apt to be misleading than helpful. Within the Carnapian framework, the parallel between the two relations is pursued in terms of ranges of propositions – full entailment involves complete inclusion of one range within another, partial entailment involves some sort of overlapping of ranges. (See Salmon,1969, for a detailed analysis of this concept.) If the measures are assigned to the ranges in an a priori way, perhaps by somehow invoking the principle of indifference, then the relation of partial entailment will embody precisely the kind of a priorism against which Venn mustered his arguments. If some hypothesis is partially entailed in a strong way by evidence, on what basis are we justified in thinking that we have a good chance of being right if we believe in it? None, Venn would say. He has argued, in effect, that if a meaningful analogy between the two sorts of entailment is to be drawn, it must be done differently. If, using the deductive relation of full entailment, we draw conclusions from true premises, such conclusions will *always* be true. If we are to introduce a probabilistic relation of partial entailment, in order to talk about the probability or degree of confirmation of hypotheses, then Venn would say (as I understand him) that we must be able to provide some basis for saying that our conclusions, based upon true premises, will *usually* be true. Relative frequency thus remains the acid test for probabilistic reasoning if such reasoning is to constitute a means for objective evaluation of beliefs. This, in my opinion, is the import of Venn's analysis of probability for the confirmation of hypotheses.

NOTE ON TEXTS: For the most part, I have referred to the first (1866) edition of *The Logic of Chance*, partly because it is historically important to consult the earliest text, but mainly because it seems to me that the discussion is more clear, crisp, and incisive than the versions found in the later and much longer editions.

University of Arizona

BIBLIOGRAPHY

Carnap, R., 1950, *Logical Foundations of Probability*, University of Chicago Press, Chicago. (2nd ed., 1962)

Hacking, I., 1975, *The Emergence of Probability*, Cambridge University Press, Cambridge.

Keynes, J., 1921, *A Treatise on Probability*, Macmillan, London.

Kneale, W. & M., *The Development of Logic*, Clarendon Press, Oxford, 1962.

Kyburg, H., et al. (eds.), 1964, *Studies in Subjective Probability*, John Wiley and Sons, New York.

Madden, E., 1960, *The Structure of Scientific Thought*, Houghton Mifflin, Boston.

Salmon, W., 1950, *John Venn's Theory of Induction*, UCLA Ph.D. dissertation.

Salmon, W., 1969, 'Partial entailment as a basis for inductive logic' in N. Rescher, ed., *Essays in Honor of Carl G. Hempel*, D. Reidel, Dordrecht.

Venn, J., 1866, *The Logic of Chance*, Macmillan, London. (2nd ed., 1876; 3rd ed., 1888; 4th ed., 1962, Chelsea, New York)

Venn, J., 1881, *Symbolic Logic*, Macmillan, London. (2nd ed., 1894)

Venn, J., 1889, *The Principles of Empirical or Inductive Logic*, Macmillan, London. (2nd ed., 1907)

von Mises, R., 1951, *Positivism*, Harvard University Press, Cambridge.

WESLEY C. SALMON

ROBERT LESLIE ELLIS AND THE FREQUENCY THEORY

John Venn was not the only nineteenth-century Cambridge man who contributed significantly to the frequency theory of probability. In 1842, Robert Leslie Ellis presented a paper to the Cambridge Philosophical Society, entitled 'On the foundations of the theory of probability', in which he deals with the relationship between probabilities and long run frequencies. Although this paper contains insights of the greatest importance, it does not seem to me to provide an adequate basis for considering Ellis rather than Venn the first frequentist. In saying this, I do not mean in the least to detract from the value of Ellis's contributions.

Let me begin by stating what I take to be his most significant ideas vis á vis the frequency theory.

(1) There is an inseparable connection between probability and long run frequency.

> If the probability of a given event be correctly determined, the event will, on a long run of trials, tend to recur with a frequency proportional to this probability.
>
> This is generally proved mathematically. It seems to me to be true a priori. (Sec. 4)

This statement could easily be taken to mean that there is an analytic or definitional connection between probability and long run frequency, and certainly Ellis hankered after some such definition, as we shall see. But in this paper he still seems content to accept the classical definition of probability as the ratio of the number of favorable to the number of equally possible outcomes. He does not claim that the long run frequency is identical to the probability, or equal to it, but only proportional. The relationship between probability and long run frequency is more complicated than simple definitional identity.

(2) Degree of expectation of a particular event is indissolubly connected with belief in the long run frequency with which events of the type in question will occur in a large series of cases.

> When on a single trial we expect one event rather than another we necessarily believe that on a series of similar trials the former event will occur more frequently than the

J. Hintikka, D. Gruender, and E. Agazzi (eds.), Pisa Conference Proceedings, Vol. II, 139–143.

> latter. The connection between these two things seems to me to be an ultimate fact, or rather, for I would not be understood to deny the possibility of further analysis – to be a fact, the evidence of which must rest upon an appeal to consciousness. Let anyone endeavour to frame a case in which he may expect one event on a single trial, and yet believe that on a series of trials another will occur more frequently; or a case in which he may be able to divest himself of the belief that the expected event will occur more frequently than any other.
>
> For myself, after giving a painful degree of attention to the point, I have been unable to sever the judgment that one event is more likely to happen than another, or that it is to be expected in preference to it, from the belief that on the long run it will occur more frequently. (Sec. 4)

This statement constitutes an admirably clear articulation of what is perhaps *the* fundamental principle of the frequentist – that probability, degree of expectation, likelihood, etc., have no meaning except that deriving from long run frequencies. The principle is, I believe, denied by probability theorists of other persuasions. The subjectivist or personalist believes that there is some feature of our belief – such as degree of conviction, or odds with which we are willing to bet – which has no necessary connection with frequency concepts at all. Such degrees of conviction can attach to the occurrence of events which are not of a type which will be repeated, and can exist without dragging in hypothetical repetitions or virtual sequences. A number of years ago, in personal conversation, I discussed precisely this issue with L. J. Savage. I attempted to defend Ellis's position; Savage maintained that the notion of the value of action – where its outcome is uncertain – made intuitively good sense to him without any consideration of actual or possible repetitions.

Like many pre-twentieth-century philosophers, Ellis puts his point in psychologistic terms. I do not believe a strong case can be made for the thesis that it is psychologically impossible for anyone to expect a particular outcome on a particular trial without believing that it will happen often in the long run. Consider the crap-shooter who just "knows" that he is going to make his point (say four) on this roll of the dice, even though he is fully aware that getting an outcome of four on tosses of a pair of standard dice is quite infrequent (3/36) in the long run. Also, one can have a strong conviction that some singular occurrence – such as the Second Coming – is about to take place, without any explicit or implicit reference to frequencies. What the frequentist must claim is that degrees of belief cannot be rational unless they are properly grounded in frequencies.

The logical theorist, similarly, maintains that there is a degree to which evidence-statements can support the assertion that a given event will occur – this is its inductive probability on that evidence – again, without reference to frequencies. I have tried to argue against this thesis elsewhere ('Partial entailment as a basis for inductive logic') and shall not repeat the arguments here. What I want to emphasize is the crucial importance of Ellis's thesis to the frequentist's position. I do not mean to suggest, of course, that it is impossible to define a relation of partial entailment on a purely a priori basis, without reference to frequencies – after all, Carnap has done so. What the frequentist must maintain, once more, is that such relations have no probative force unless they are appropriately related to frequencies. Lacking such a relation to frequencies, the relation of partial entailment cannot serve as a foundation for a concept of degree of rational belief.

Contemporary single-case (as opposed to virtual sequence) propensity theorists believe that tendencies exist which have no necessary connection with frequencies. As I understand them, such propensity theorists hold that long run frequencies probably match the propensity in value, but not necessarily. Whether such a single-case propensity interpretation can be carried through satisfactorily is, I believe, an important current question.

With the benefit of hindsight, it might be easy to construe Ellis's remarks as the claim that probabilities associated with single events must be defined or explicated in terms of long run frequencies. However, Ellis does not come right out and say so; instead, he applies the foregoing considerations to the classical interpretation:

> It follows as a limiting case that when we expect two events equally, we believe they will recur equally on the long run. In this belief we may of course be mistaken; if we are we are wrong in expecting the two events equally, and in thinking them equally possible. Conversely, if the events are truly equally possible, they really will tend to recur equally on a series of trials. (Sec. 5)

What Ellis does in this passage is to take the classical definition of probability in terms of equally possible cases, and to explicate equal possibility in terms of equal long run frequency, instead of, with Laplace, in terms of absence of reasons to have unequal expectations. Ellis thus attempts to provide an objective version of the classical interpretation, in contrast to the usual subjective construal.

(3) Ellis attacks the notion – to be found in many discussions of the

Bernoulli theorem – that mere a priori counting of possible cases can be informative concerning actual probabilities. Given that there are many more possible ways in which 100 tosses of a coin can occur in which 50 outcomes are heads than in which 100 heads occur, it still does not follow that 50 heads and 50 tails is more probable than 100 heads, for it is not a priori impossible that, on performing the experiment many times, 100 heads would occur more often than the 50;50 result.

But, although a series of 100 heads can occur in one way only, and one of 50 heads and 50 reverses in a great many, there is not the shadow of a reason for saying that therefore the former series is a rare and remarkable event, and the latter, comparatively at least, an ordinary one.

Non constat, but the single case producing 100 heads may occur so much oftener than any case which produces 50 only, that a series of 100 heads may be a very common occurrence, and one of 50 heads and 50 reverses may be a curious anomaly. (Sec. 7)

Ellis is here concerned to undermine the use of the Bernoulli theorem as a means to prove a priori that the actual frequency of occurrences matches their probability, ascertained on some a priori basis. His criticism of such uses of the probability calculus are, of course, valid. They constitute one important plank in the frequentist's platform.

My general assessment of Ellis's contribution is that he furnished several significant steps toward the frequentist position, but he did not succeed in clearly articulating that theory. At the end of his paper he adds, "With a view to conciseness I omit several interesting points which here present themselves – namely, . . . what seems to me to be the true definition of probability, founded on a reference to the ratios developed on the long run." But he does not – by his own explicit admission – discuss that point.

When, in 1854, Ellis presented additional 'Remarks on the fundamental principle of the theory of probabilities', he did not fill this particular gap. He offers some metaphysical considerations to support the thesis, "On a long run of similar trials, every possible event tends ultimately to recur in a definite ratio of frequency", which is the fundamental principle. He hardly succeeds in clarifying the meaning of his fundamental principle in the following remarks:

. . . it is less obvious how the idea expressed by a "long series of trials" in "definite series of experiments" and the like, is to be expressed, so as to make the analogy between the fundamental principle of the theory of probabilities and those of the other sciences more obvious than it has hitherto been. The idea in question is not readily expressed in any way, because in its own nature it is negative and indefinite. The phrases I have just quoted imply merely the absence of the limitations inseparable from individual cases, or from any finite number of such cases, whether contemplated as actually existent or as about to be developed within definite limits of space and time.

Such remarks fail, among other things, to enable us to understand how "ratios developed on the long run" can be explicated in terms of the limit of the relative frequency in an infinite sequence of events.

What did Venn do which Ellis failed to do? (1) Venn abandoned altogether the classical conception of probability in terms of the ratio of the number of favorable to equally possible cases. He realized that equiprobability is a totally dispensable notion so far as the fundamental definition of probability is concerned. (2) Venn explicitly recognized that probability has to do with aggregate phenomena, and that if a probability concept is to be applied to single events, it must be derived from the long run frequency. Moreover – and this is the crucial point – he developed an explicit theory of the probability of the single case within the context of a frequency theory. Ellis (in retaining a version of the classical interpretation) did not relinquish the view that probability is directly applicable to single cases, without derivation from a more basic concept. (3) Venn explicitly introduced the notion of the limit of the relative frequency (in the technical mathematical sense), and discussed with considerable care the relationship between the abstract mathematical conception of an infinite sequence and the classes of physical entities with which we deal in the real world. (4) Venn dealt more explicitly than did Ellis with problems relating to the probability of hypotheses, and by virtue of this fact, he developed a frequency theory of probability which comprehends both the use of probability concepts in dealing with statistical aggregates and the use of probability concepts in connection with questions involving the weight of evidence.

Ellis, it seems to me, took us to the very threshold of a frequency theory of probability; Venn opened the door and led us in. I have no desire to say which was the greater accomplishment. Both were superb.

[Terminological note: In this discussion, the term "frequency theory" is used in the sense explained in 'John Venn's *Logic of Chance*', where it is explicitly distinguished from the term "frequency interpretation".]

University of Arizona

SECTION V

THERMODYNAMICS AND PHYSICAL REALITY

LORENZ KRÜGER

REDUCTION AS A PROBLEM

Some Remarks on the History of Statistical Mechanics from a Philosophical Point of View

1. ONTOLOGICAL AND EPISTEMOLOGICAL ASPECTS OF REDUCTION

The endeavour to develop a unified world view which systematizes and explains the heterogeneous mass of our experiences has been an important motive in scientific research since its early beginnings. It involves, among other things, the attempt to "reduce" certain phenomena to others that are qualitatively different. Thus, an account of the world that speaks only about bodies and their motions appears to be preferable to an account that comprizes bodies, motions, and, in addition, phenomena like heat and cold as irreducible or unexplained items.

But not only is it a goal to simplify or minimize the ontology of our world view, it is equally important to exhibit the proper ontological dependencies. Thus, heat and cold are understood to be qualities of things which already presuppose the existence of bodies, whereas spatial and temporal properties, including motion, appear to be constitutive for the very notion of body. Intuitively, at least, it seems clear that it would be out of the question to reduce the phenomena of visible motion to heat and cold, whereas the reverse reduction is easily conceivable. In short, reduction programmes are shaped by ontological preconceptions. Conversely, these preconceptions will be expected to be born out by the realization of those programmes.

To establish the reductive connections between qualitatively different domains, *e.g.* between heat and motion, implies, among other things, the need to allow for the possibility of saying that something which, in our experience, does not have a certain property or is not of a certain kind, nevertheless has this property or is of this kind, *e.g.* that heat, which is entirely different and independent from motion in our experience, is eventually nothing more than motion. Reduction, therefore, cannot be achieved without involving hypotheses about unobservable entities. Thus, the attempt at reducing heat to motion

J. Hintikka, D. Gruender, and E. Agazzi (eds.), Pisa Conference Proceedings, Vol. II, 147–174.

appeals to hypotheses concerning the structure of matter and, in doing so, goes far beyond what is immediately warranted by the phenomena to be reduced. From the indefinite set of conceivable models which are compatible with the ontological programme, a choice must be made, *e.g.* in favour of atomism, of a certain structure of atoms, of certain laws of interaction *etc.*

This means that the execution of a programme of *reduction* will be interwoven with the *exploration* of "deeper" levels of reality beyond the level of our experience. This state of affairs points to an important epistemological aspect of reduction: it is no less a tool of investigation than it is of explanation. Fruitful reductionist assumptions may support certain views of the structure of things, and the previous and independent knowledge of this structure could justify certain reductionist claims. To what extent these two aspects of reduction can be separated in actual scientific practice is an open question; the answer may essentially vary with the concrete case at hand. It is clear, though, that the epistemological dependence of different domains tends to be counterdirected to the ontological dependence of the same domains: the knowledge of macrophenomena may be used for justifying views of microphenomena that, in their turn, are taken to reduce these macrophenomena.

In scientific practice, this mutual dependence is likely to exclude the *effective* elimination of theories that are (or are asserted to be) reduced. Nevertheless, the reduced domain may be considered to be eliminated from an *ontological* account of the world, in the sense that it is shown to consist of mere epiphenomena produced by the "primary" entities and properties. Yet this interpretation will only be possible as long as the counterdirected epistemological dependence does not become manifest in the reductive account, *i.e.* as long as the latter does not contain uneliminable epistemic terms that are needed in order to make the reduction possible in the first place. It remains to be seen whether or not this requirement can be satisfied in the case of microreductions. The reduction of heat to motion is a prominent test case which deserves a re-examination with questions like this one in mind.

2. REDUCTION AS AN INTERTHEORETIC PROBLEM

In most accounts of reduction the answer to those questions is prejudged. Reduction is conceived of as the *deduction* of the theory to be reduced, T_A, or a theory T_A^* that closely approximates T_A, from the reducing theory, T_B.[1] The items defined on the basis of T_B and the statements about them deduced from T_B may then, as indicated above, be interpreted as epiphenomena, and in this sense as eliminable from the primary picture of the world which is given in terms of T_B alone.

Since it is a characteristic fact of actual science that it offers relatively independent theories for different domains of things or experiences, it is certainly adequate to conceive reduction as a problem in intertheoretic relations.[2] It is less clear, though, whether it will always be appropriate to describe the relations involved as derivability relations, so that the reductions of T_A to T_B can be viewed as an "absorption" of T_A into T_B.[3] Modified or alternative accounts seem to be worth investigating. The absorption may, for instance, be effected into a revised theory T_B^* rather than into T_B;[4] and this case may be significantly different, depending on the extent to which the adjustments of the reducing theory that transform it from T_B into T_B^* are determined by the very reduction of T_A in question. A more widely diverging alternative account construes reduction as the development of a new "interfield" theory comprizing elements from both T_A and T_B.[5] One might even consider the possibility that some so-called reductions are not even fully expressible in terms of both object theories involved, but require an additional account in terms of suitable metatheories (*e.g.* metaphysical, epistemological, historical, or pragmatical in character) that describe, analyse and explain what is involved in the simultaneous use of the theories to be related.[6]

The ontological import of reduction will, of course, strongly depend on what concept or concepts of reduction prove to be adequate to actual scientific cases. Whatever the outcome, the articulation of the originally ontological programme of reduction in terms of intertheoretic relations is intended to do justice to the advice, given by Nagel and Hempel in this context, to the effect that ontological issues should be spelled out in concrete linguistic terms.[7] The ontological problem of reduction is then that of an empirically meaningful (and hopefully adequate) unified theoretical language about the world, or a particular domain of the world.

The evaluation of the ontological achievement will, accordingly, proceed by tracing the sources and interpretations of the terms and statements that occur in the unified theoretical account which is produced in the reductive attempt. Only if this account can be shown to be free from elements that belong to the theory to be reduced and free from epistemic elements that disturb the coherent presentation of the reducing realm of entities and phenomena, will it be justified to draw conclusions of the form "*A* is nothing more than *B*", *e.g.* heat is nothing more than motion.

3. REDUCTION AND PROBABILITY

It has already been mentioned that the reduction of a domain *A* to another domain *B* which is qualitatively different from the first, requires particular assumptions as to how it is possible that the things in *A* do not show properties typical for *B*, although they eventually turn out to have them. The easiest way to make room for an explanation of this possibility is to assign the different kinds of properties to different orders of magnitude, *e.g.* continuous change to the macrolevel of common experience, and discreteness and stability to the microlevel of atoms; similarly thermal properties to the former, and the motions which they are asserted to result from, to the latter.

Now, it is obvious that individual microprocesses are either unobservable, or, at best, only observable via certain magnifying mass processes which involve a large number of individual microprocesses. It is, therefore, understandable that statistical ideas entered into physical theory as soon as the microlevel processes became a serious object of research and reductive explanation. There are a number of studies which analyse how the need to deal with large numbers of systems in the kinetic theory of matter led to the inclusion of the concept of probability into the *object* theory of physics, as opposed to merely auxiliary *methodological* theories like error theory, or *meta*theories like those of inductive reasoning.[8]

Thus it is an important feature of the reduction of heat to motion with respect to the interpretation of thermodynamics in terms of mechanics that it involves the notion of probability. This notion has been ambivalent as to its interpretation from early on;[9] and it remains an open question today whether it is to be understood as "subjective" or "epistemological" (in some broad sense comprizing a number of finer alternatives), or rather as "objective" or "ontological" (again in a broad sense), or, finally, whether different though related concepts are required.[10]

It is to be expected that this problematic situation will affect reductions which involve the concept of probability. In case we are, for instance, inclined towards an epistemic interpretation, we are led to ask whether that interpretation might not be incompatible with the ontological aim of the reduction, *i.e.* the attempt at a unified account of both domains involved. Or is, rather, the opposite true of the epistemic view: to wit, that it makes the introduction of probability irrelevant and innocuous to the ontological reduction programme? If not, cannot probability be given an ontological and objective physical interpretation and explanation in terms of the reducing theory, *i.e.*, in our test case, in terms of classical particle mechanics? Again, if not, are there no ways to eliminate or circumvent probabilistic considerations after all? These questions (and others like them) suggest themselves readily when one begins to study the history of the reduction of thermodynamics to mechanics more closely.

It is the leading idea of the rest of this paper that all serious problems connected with this paradigm case of a microreduction are tied to the introduction of probability. The revolutionary character of this step and the difficulties connected with it were, for a considerable time, not, or only insufficiently, recognized by physicists. A number of reasons may be adduced for this unawareness: (1) probability was taken as an *a priori* mathematical concept that is sufficiently clear in itself and obvious in its applications; (2) the physical explanation of stable averages, randomness, *etc.* was believed to be unproblematic; (3) probability was viewed as no more than a tool of representing our limited knowledge (our ignorance), and therefore immaterial to the objective picture implied or intended by our representation.

That all three reasons actually contributed to a serious underestimation, if not a misapprehension, of the problem is born out by the history of the case. As long as reason (1) was prevailing, no need could be felt to distinguish, let alone decide, between a subjective and an objective interpretation of probability. Even after the empirical import of probabilistic considerations had been recognized, a serious discussion of the interpretation of probability did not arise among physicists, and the status of this concept in physics remained obscure. A decision between objective and subjective views was certainly not reached, and widely not even conceived as an issue during the entire development of (classical) statistical mechanics.[11]

Instead, competing objective explanations of probabilistic phenomena were considered: (1) the assumption of unobserved deterministic processes which resulted in a randomization of molecular processes, (2) the assumption of contingent cosmological boundary conditions, and (3) the assumption of irreducibly indeterministic dynamics, though this last alternative remained marginal until the rise of quantum mechanics, and will, therefore, not be considered any further in the sequel.[12] The possibility of explaining probability along the lines of reason (2) tended to conceal the need to explore this notion any further. On the other side, the unsatisfactory character of these explanations was (and still is) a reason to increasingly favor the epistemological line of type (3) above.

Yet even after the formulation of present-day (equilibrium) statistical mechanics had been achieved by Gibbs and Einstein, the full import of introducing probability into the object theory of physics was not recognized. Einstein offers, perhaps, the most prominent instance for this statement. He based his mechanical theory of thermodynamics on the ensemble approach, and he used the term 'probability' at crucial points of his argument without bothering to comment on it. It is then, perhaps, not surprising that he made unqualified reductionist claims like "*der zweite Hauptsatz erscheint also als notwendige Folge des mechanischen Weltbildes*"[13], although only a few pages earlier he had described the task before him as that of "*die Sätze über das Wärmegleichgewicht und den zweiten Hauptsatz unter alleiniger Benutzung der mechanischen Gleichungen und der Wahrscheinlichkeitsrechnung herzuleiten*".[14] His life-long resistance against the recognition of probabilistic traits in reality itself is well known, so that he can only be understood to have taken for granted that the usual probabilistic accounts were only a matter of convenient practice, but not of principle. What he did not see is that this confidence in the principle remained without warrant in all actual arguments. The unquestioned coexistence of statistical theories with a deterministic world view appears to be typical for the entire prequantum probabilistic period of physics.

The assertions contained in this section will now be substantiated in the following two sections.

4. 'PROBABILITY' IN BOLTZMANN: *A PRIORI* AND ONTOLOGICAL TENDENCIES

4.1. *Boltzmann before* 1877

As far as I can see, there is no noticeable change in Boltzmann's views on probability, or in the effective use he makes of this concept up to 1875. I feel, therefore, entitled to evaluate his earlier writings together, without paying attention to any development of his thought concerning this topic.

It has been doubted that Boltzmann ever thought in truly probabilistic terms before he was confronted with Loschmidt's objection to his H-theorem.[15] This was certainly not his own view of the matter. Daub[16] has pointed out that Boltzmann was right when in 1877[17] he claimed to have clearly said already in 1871 that a mechanical proof of the entropy law for non-periodic molecular orbits "is *only* possible through consideration of the probability of the various atomic positions".[18] And in 1872 he said: "The problems of the mechanical theory of heat are . . . problems of the calculus of probabilities." The reason, though, he offers for this view indicates a fairly limited conception of probability: "The determination of average values is a task of the calculus of probabilities."[19] Klein rightly emphasizes that Boltzmann in 1872 did not yet see that the H-theorem itself is of a probabilistic character. What concept of probability, then, did Boltzmann have in his earlier writings?

Klein[20] distinguishes "two quite different meanings" that Boltzmann attaches to the distribution function f, which describes the state of a gas and which – suitably normalized – is the formal expression for the probability of finding a particular molecule in a particular phase: (i) f is interpreted as that fraction of a long interval of time during which a single molecule is in a given state: (ii) f is interpreted as the fraction of the total number of those molecules in the gas which are in a given state. According to Klein, Boltzmann soon realized that, in view of this situation, "a definite hypothesis was called for", the ergodic hypothesis, first found (in a rather implicit version) at the end of the very paper that starts with the dual interpretation of f[21] and more explicitly introduced in 1871.[22] Did Boltzmann attach also two "meanings" to the term 'probability' itself? And did he introduce the ergodic hypothesis in order to relate them to each other, which is indeed a job usually attributed to this hypothesis? I suggest that the answer to both questions be 'no'.

Klein is certainly right in drawing attention to Boltzmann's two ways of interpreting the distribution function, and hence also the concept of probability. The double view is pervasive all through the papers of gas theory. A particularly clear passage is found in the H-theorem paper of 1872: "... first of all we have to determine the probability of the various states that are found in one and the same molecule during a very long time and in the different molecules at the same time".[23] But this passage in itself, as well as many others, renders it doubtful that one should speak of two different "meanings" here. It seems more appropriate to talk about two different applications, or perhaps two different ways of determining numerical values of probabilities. Boltzmann never even hints to the need of explicitly defining a meaning of 'probability' for physics. Otherwise he could hardly have introduced f as the number of molecules in a given state without any comment. The absence of a justification or an explication rather points to the self-evidence of the classical meaning: probability as the number of favourable cases divided by the number of all possible cases. For the physicist, the cases could well be the molecules of a given gas at a given time.

This reading is supported by two further observations. First, in view of the classical concept of probability it is less obvious how one could speak of the probability for only one system (molecule). What are the distinguishable and countable cases which are needed in order to obtain a numerical value? It is here that Boltzmann turns to the idea of the fraction of time. This answer could seem obvious to the physicist; and indeed Boltzmann does not digress into anything like an explicit justification of it. When he first introduced the probability of a given velocity c of a single molecule $\psi(c)\,dc$ in 1868 he simply added an explication for this magnitude $\varphi(c)\,dc$: "the sum of the time moments during which through a very long time the velocity ... lies between c and $c + dc$". In the opening passage of the 1868 memoir he continued with an explicit gloss of the form, "hence the probability that c lies between c and $c + dc$".[24] On the other hand, the talk about the number of systems per area or volume is, as far as I am aware, never accompanied by such a comment. At most, Boltzmann adds an attribute like 'average' ("*durchschnittlich*") to 'number'. Second, Boltzmann frequently based probability claims on *a priori* considerations from indifference – a typical feature of classical probability theory. Several of the memoirs on gas theory from 1866 through 1872 begin with restricting assumptions to the effect that the velocity distribution is uniform in space, or that all velocity components are equiprobable. Typically, Boltzmann just states that this much is "clear" or "self-evident" ("*von selbst klar*", "*selbstverständlich*", "*ganz in der Natur der Sache liegende Beschränkungen*" *etc.*).[25] That *a priori* indifference considerations are really at the basis of these equiprobability claims is born out by the beginning of the 1875 memoir on thermal equilibrium under external forces; with respect to these forces, Boltzmann explicitly disclaims his by now nearly habitual simplifications.[26]

From all this, I conclude that Boltzmann, at least before 1877, never felt any need to reflect about the *meanings* of probability. He most probably had the classical definition in mind; he even used the classical argument from equal probabilities quite freely. The natural physical interpretation for him was in terms of relative frequencies. Only in the special case of single systems had he to resort to the peculiar device of interpreting them (*not*, of course, computing them) in terms of "time moments".

Boltzmann obviously saw no need of special arguments to connect one method of interpreting or of determining a relative frequency (relative number of systems) with another such method (fraction of time). In particular, the ergodic hypothesis does not have this function for the early Boltzmann. The first claim is instantiated by a typical procedure used in many of his papers[27]: the fraction of time during which a given molecule has a velocity between c and $c + dc$, $\varphi(c)\,dc$, is simply multiplied by the average number of systems N per unit area or volume, thereby furnishing a new average number. The two types of probability magnitudes are used on an equal footing without the least gloss. True, usually Boltzmann spoke only of time fractions as "probabilities", whereas distribution functions over many systems are often simply called "(average) numbers". But the average number simply didn't need the commentary: it was clearly a probabilistic magnitude or a relative frequency, whereas it needed explaining that a time interval can be viewed in the same way.[28]

It would therefore be misleading to say that, for Boltzmann, it was part of the meaning of probability that it was a magnitude defined (or primarily defined) for *time* ensembles. His conception of probability as a time fraction is no more probabilistic in the genuine sense than that for numbers. Moreover he could, of course, make no *effective* use of the time fraction idea at all, since there is no way to follow up the path of a single molecule, but he had always to resort to distribution functions, whatever their interpretation. And he moved from one interpretation to the other without giving the least thought to the transition.[29] It seems significant also that he never attempted to interpret the (average) numbers with respect to time fractions in the history of the system. This thought made its appearance only in the 1890's when he turned to a discussion of the overall time dependence of entropy.

Furthermore, it should be noted that Boltzmann does not appeal to the ergodic hypothesis at any of the numerous places where he moves forth and back between the two interpretations of the distribution function.[30] What, then, is the function of the ergodic hypothesis? Since this question is of no immediate concern here, it suffices to indicate that there *are* clearly recognizable purposes attached to it which are different from establishing a relationship between time ensembles and space ensembles. One is that of circumventing the often forbiddingly long derivations within the framework of collision theory[31]; and a second, and more important, one is that of generalizing results beyond the limits of particular models that Boltzmann had been able to deal with explicitly so far.[32]

A further decisive reason against the assumption that for Boltzmann the ergodic hypothesis had anything to do with the clarification or justification of the concept of probability in physics derives from the fact that Boltzmann always believed one could, and in the end must, do without the "hypothesis". After having introduced it explicitly in his second paper of 1871 on gas theory (1871 b), he hastened to leave it behind again in the next paper (1871 c); and, as is well known, he insisted in 1872 that he had given "a strict proof", which was "free from all hypotheses".[33] And this did not mean for him: free from probability considerations, as he is eager to point out explicitly.[34] He tended to circumvent the hypothesis all through his life; certainly he did not think of it as something "called for".

That leaves us with the question what, if any, justification he had for the use of probability in the derivation of what he, for a time, believed to be a strict law of nature derived in a rigorous demonstration. It is quite clear that he did not conceive of the distribution function as of something epistemic referring either to our knowledge or to the inaccuracy of our measuring devices.[35] With respect to his, as it were, naive use of probabilistic terms and his tendency to lose sight of what little truly probabilistic content they had, it is not surprising that we find no more than a few hints. Yet these, at least, seem clear in themselves; moreover they do service for explaining the possibility of adequacy of applying both concepts like 'probability', 'average', *etc.* in general, and the ergodic hypothesis in particular. Boltzmann, first, refers simply to our experience of the stability and uniqueness of thermal equilibria and of the irrelevance of initial phases of the motion of bodies for their behaviour under thermal contact.[36] Second, and more important: he refers to "the multiplicity of forces that act on the bodies from outside",[37] or to the "quite uncalculable cooperation of the most variegated external conditions" in order to explain the complete irregularity of thermal motion and the stability of the observed averages.[38]

The first observation does not have any justificational or explanatory ring: in one of its two occurrences it is accompanied by an additional *a priori* reason, whereas in the second it is immediately followed by an explanation from the nature of the objects concerned. On the other hand, the second observation appears to express Boltzmann's background convictions in accordance with his well known deterministic inclinations, which for quite some time prevented him from recognizing the statistical nature of his H-theorem.

In view of the attempted reduction of thermodynamics to mechanics, one should point out that the explanation of irregularity by the superposition of many external factors only pushes the problem one step further. Moreover, if this explanation should be correct, it would have been quite unnecessary to bother with the exceptions to the second law that Boltzmann discussed in 1875. His final solution does, indeed, not appear to have relied on this argument, but rather on cosmological assumptions (cp. sect. 4.3 below).

4.2. *Boltzmann in* 1877

The two papers of 1877 in which Boltzmann answered Loschmidt's objection to his H-theorem, and developed an entirely independent new proof of it, have received a careful analysis[39], so that I shall refrain from going into the details. For my purpose it will suffice to recall a few main points here.

Upon the impact of Loschmidt's argument, Boltzmann now declared that there is no way of proving "with absolute necessity" that a gas will reach a uniform distribution even in the long run. "This is already taught by the calculus of probability itself".[40] The statement appears to revoke explicitly what he had said in 1872:

"It would be a mistake to believe that heat theory suffers from uncertainties, because the theorems of the calculus of probability are applied to it. One should not take a theorem of the calculus of probability that has been fully demonstrated for a theorem whose proof is incomplete . . . ".[41]

On the other hand Boltzmann continues to call his derivation of the H-theorem an "exact proof", which shows that H "*can only* decrease".[42] Understandably, Klein found Boltzmann's insistence in 1877 that he had pointed to the probability aspect of his mechanical theory of heat ever since 1871, "somewhat disingenuous, or at least, misleading".[43] I think we should conclude from the apparent incoherence of Boltzmann's statements that he considered the possible exceptions to the second law as a merely mathematical feature of the theory that would never become relevant in reality. Perhaps he thought that disturbances from outside would always destroy the highly artificial arrangements that are necessary for the survival of the exceptional systems.[44]

That his basic commitments had not changed very much may also be said with respect to the idea of probability itself: the fundamental change in the derivation of thermodynamic properties is not accompanied by a comparable change in the concept. It remains closely related to the classical definition. The whole calculation of permutabilities approaches an *a priori* consideration: to what degree are specific macrostates possible? There is not even an allusion to something dynamical like time fractions; the model of computation is that of an urn filled with slips on which energy or velocity values are noted. All the physics goes into the discussion of which of the two numberings is appropriate with respect to the laws of mechanics, and into the (ergodic-like) constraints under which the most probable state of a gas is determined: to wit, for fixed particle number and fixed total energy.

What is even more significant is the complete lack of any dynamical consideration at that place where one would expect it most: *i.e.* in the treatment of the development of a system through time, which had, moreover, been the topic Loschmidt's objection was concerned with. We only find the often quoted sentences:

"In most cases the initial state will be very improbable; the system passes from this through ever more probable states, reaching finally the most probable state, . . . ".[45]

Does this statement imply that Boltzmann had adopted a stochastic view of the fundamental dynamics?[46] That could hardly be the case, since at the same time he relied so heavily on Liouville's theorem. Boltzmann could not consistently believe in the deterministic basis of his earlier and his present derivations of the second law and simultaneously claim any explanatory power for the sentence just quoted.[47] It seems significant to me that, later on, he did not make very much of his new treatment of the second law; he finally subsumed a brief sketch of it under the (only seemingly misleading) title "the *mathematical* meaning of H".[48]

How could he then think he had defended his reduction of the second law to mechanics? We have already seen that he tended to push the problem one step further in assuming irregular disturbances from outside. Yet a second wayout would seem to suggest itself: to make special assumptions about the initial states that really occur in nature. The description of the "dynamics" quoted in the last paragraph seems to confirm this assumption. But a few months earlier Boltzmann had already explicitly argued that it would be a "*non-sequitur*" and no more than an "interesting sophism" to conclude from Loschmidt's objection that the second law is not founded in the basic laws, but merely in the initial conditions.[49] This reluctance to embrace an obvious solution is especially surprising in view of the fact that only a few pages later Boltzmann discusses the application of Loschmidt's objection based on the past: if it allows us to conceive of processes with decreasing entropy at all, then it does this also for systems in the past. We could then imagine that the present low entropy state of a system emerged from states of higher entropy. Exactly the same argument suggests itself *a fortiori* when one thinks in terms of probability measures. Apart from assumptions about initial conditions, one would have to assume states of the world that are more probable than ours in the past.[50] The result is that in 1877 we see Boltzmann further than ever from accomplishing an ontological reduction of thermodynamics to mechanics, though he does not seem to have been aware of this.

4.3. *Boltzmann on H-curves*

All ingredients of Boltzmann's thinking on probability that have been mentioned so far continue to be present through the rest of his life. I do not want to pursue them further or balance them with new thoughts on the subject; the most they can contribute to the (ontological) reduction of heat to motion is to shift the problem to background assumptions. These concerned irregular disturbances from outside which could not be included in the scope of the available theories, but remained unexplained presuppositions of their use.[51]

In view of Poincaré's recurrence theorem, the only remaining hope for achieving the reduction within the deterministic framework seems to be connected with the question of contingent initial conditions that obtain in our part of the world. Boltzmann could finally not fail to realize that, after he had been confronted with Poincaré's result by Zermelo.[52]

Since the exchange with Zermelo is well known and has been carefully analysed[53], I shall only comment on the significance of the final outcome for the question of reduction. Already before the recurrence objection became known to Boltzmann, he had discussed the H-curve in order to clarify the content of his by now famous theorem.[54] In this context he also for the first time introduced a cosmological speculation which he attributed to his assistant Dr. Schuetz: our part of the world with its remarkably high order or H-value had emerged by a chance fluctuation from a universe in thermal equilibrium. The probability of this occurrence could be made as high as one liked, by simply assuming a correspondingly large universe. Though this idea was highly speculative, it could serve to show the compatibility of an H-curve that is time-symmetric, on the average, with a state of our world that develops almost inevitably towards higher entropy. Yet the argument was still insufficient for explaining our beliefs about the past, *i.e.* that we usually interpret present systems of low entropy to have evolved from states of still lower entropy. In other words, we assume ourselves to be placed on a decreasing branch of the H-curve. Perhaps Boltzmann just forgot to point out this additional assumption; perhaps he still thought, as he had in 1877, that one should base the whole case on probability theory and not resort to special hypotheses concerning initial states. When he wrote his first answer to Zermelo in 1896, Boltzmann had realized that more was needed: *viz.* the assumption "that the complex of bodies observed by us happens to be in a very improbable state".[55] A little later he explicitly contradicts his earlier view of 1877 quoted above[56]:

"We proved that [the magnitude H] continually decreases as a result of the motion of the gas molecules among each other. The one-sidedness of this process is clearly not based on the equations of motion of the molecules. For these do not change when the time changes its sign. This one-sidedness rather lies *uniquely and solely in the initial conditions.*"[57]

This is a belated homage to Loschmidt as well as a concession to the effect that *probability considerations are by themselves insufficient for understanding the second law.* In his second reply to Zermelo, Boltzmann had very clearly pointed out what is at stake:

"*Der Zweite Hauptsatz wird mechanisch durch die natürlich unbeweisbare Annahme A erklärt, dass das Universum, wenn man es als mechanisches System auffasst, oder wenigstens ein sehr ausgedehnter, uns umgebender Teil desselben von einem sehr unwahrscheinlichen Zustand ausging und sich noch in einem solchen befindet.*"[58]

This reduces the mechanical derivation of the second law to a bare postulate.[59]

What is still more disappointing for the hopeful reductionist is the complete lack of a mechanical analysis of the behaviour of H-curves. Boltzmann's arguments are all based on purely *a priori* probability considerations; his models for illustrating the time behaviour of *H* are games of dice and of drawing black and white balls from an urn.[60] In this respect he has not advanced beyond his position in 1877. Most likely, Boltzmann did not intend those considerations to be more than illustrations. As Brush has pointed out[61], no evidence was presented for the claims about the H-curve other than the original H-theorem. But the latter did *not* lead to a time-symmetric H. The best one can say is expressed by the Ehrenfests[62]: on the one hand, Boltzmann was right in claiming that no incompatibility had been demonstrated in his mechanical account of the second law, but on the other hand, a consistency proof had not been given either – it would involve a (presumably purely mechanical) proof of the properties of the H-curve.

In other words: the case of cosmological initial conditions had not been advanced much further than that of external disturbances: Both were purely defensive. No contradiction had been advanced that threatened the attemped ontological reduction. But the reduction had not effectively performed either. It remained a programme, though a highly attractive one. The *actual* progress that had been achieved, though, did not consist in a comprehensive mechanization of the physical world picture, but rather in a fruitful cooperation (possibly including mutual criticism) of qualitatively different theories.

It is interesting to see how Boltzmann, somewhat defensively, turns his thoughts to these incontestable scientific advantages. In his second rejoinder to Zermelo he concedes that it was "a matter of taste" to decide whether one would be willing to pursue cosmological speculations (of the type Boltzmann could *not* dispense with). But it was not a matter of taste, he continued, to choose between the Carnot-Clausius thermodynamics on the one side and the mechanical picture on the other. The first theory is undeniably the simplest expression of the facts, hence important. Yet the second fits all experiences, too, and moreover points to the possibility of novel experiences. Boltzmann concludes by pleading for the "*Allseitigkeit unserer Gedankenbilder*", that is for the combined study of thermodynamics *and* mechanics.[63] The same point is taken up again a year later in the second volume of his "*Gastheorie*". The following passages may be worth quoting:

"*Wie die Differentialgleichungen bloß eine mathematische Rechenmethode darstellen, deren klare Bedeutung nur durch Bilder erfaßt werden kann, welche von einer großen endlichen Zahl von Elementen ausgehen, so ist* ***neben*** *der allgemeinen Thermodynamik und, unbeschadet ihrer* ***nie zu erschütternden Wichtigkeit, auch*** *die Pflege mechanischer Bilder zu ihrer Versinnlichung für die Vertiefung unserer Naturerkenntnis förderlich und zwar nicht trotzdem, sondern gerade weil sie sich nicht in allen Stücken mit der allgemeinen Thermodynamik deckt, sondern den Ausblick auf neue Gesichtspunkte bietet.*".[64]

There can be no doubt where Boltzmann's heart is, but neither can there be a doubt that he appreciated the good reasons for moderating his claims, at least within the context of a scientific discourse.

5. PROBABILITY IN THE ENSEMBLE APPROACH: EPISTEMOLOGICAL TENDENCIES AND COMPLEX PERSPECTIVES

5.1. *Maxwell*

In 1878, Maxwell adopts Boltzmann's ergodic hypothesis (as presented in 1868) and makes it the basis for his general proof of his velocity distribution in the thermal equilibrium. He finds it plausible in view of *external disturbances* to which all systems seem to be exposed. In distinction to Boltzmann, he used the hypothesis in order to justify the microcanonical ensemble. The latter, though, is for him nothing but a "convenient" form of presentation, whose entire physical content and justification stems from the deterministic laws of motion plus ergodicity.[65]

As to the interpretation of probability, Maxwell is even less explicit than Boltzmann; the meaning of the concept appears to have been no problem for him. Boltzmann once states that Maxwell had determined probability in terms of ensembles, whereas he himself had preferred the characterization in terms of time fractions.[66] In reality, though, Boltzmann had of course not been able to use the time fraction idea in any of his arguments, nor had he anywhere attempted to connect these arguments to this idea. Yet he had, indeed, never accepted the ensemble approach (though he had adopted it occasionally as a preliminary shortcut for results which, in the end, would have to be obtained by collision theory). On the other hand, the ensemble approach is taken by Maxwell to be nothing more than a convenient reformulation of the assumed ergodic behaviour of thermal systems; and the ergodic hypothesis was the only genuine application of the time fraction idea of probability to statistical mechanics.

In view of this fact, it may rightly be claimed that Maxwell did achieve an ontological reduction, but only under the presupposition of ergodicity. Since that, again, was grounded in the disturbance assumption, he shifted the problem to the background surrounding the systems which were the proper objects of the theories involved in the reduction problem. Eventually, thus, he did not advance further than Boltzmann, even apart from the fact that his treatment did not cover irreversible processes at all.

5.2. *Bryan and Burbury in* 1895

The ensemble approach did not flourish after Maxwell's death, until Gibbs adopted it in 1902, presumably because of its abstract and (apart from the ergodic hypothesis) quite unphysical character. Maxwell had not suggested the possibility of an epistemological interpretation of the ensemble, and ergodicity could well seem pretty dubious an assumption: Rayleigh called it an "assumption ... of a severely restrictive character".[67]

Somewhat surprisingly, an epistemic turn shows up in the midst of collision theory. In a somewhat operationalist vein, G. H. Bryan and S. H. Burbury argued that thermodynamics applies to systems which are brought into contact with a refrigerator or a reservoir, and that this process may be thought to bring about Burbury's "condition *A*" which was needed for the usual treatment of collisions.[68] Thus, the compatibility of thermodynamics and mechanics is here secured by human interference. Yet, this would seem to frustrate the attempt at a straightforward *ontological* reduction (cp. 5.5 below).

5.3. *Jeans*

Jeans[69] took a further step toward an epistemological interpretation of probability, though he, his topic being gases, still based most of his work on collision theory and the assumptions which were usually made in that theory. The overall outlook of his work is predominantly Boltzmannian, yet with an *epistemic* twist in the scattered foundational remarks.

In the basic chapters of the book he juxtaposes the "method of collisions" and the "method of statistical mechanics" and proceeds to a comparison of them.[70] The first method requires approximations of limited validity, like the *Stoßzahlansatz*. But such assumptions, he thinks, are still preferable to Maxwell's procedure of founding statistical mechanics on the ergodic hypothesis, because this hypothesis is dubious, and so is Maxwell's reasoning concerning outside disturbances that was meant to support it.[71]

The true foundations of the statistical method had, therefore, to be sought elsewhere; and Jeans essentially adopts Boltzmann's approach of 1877. Yet he adds his own interpretation of the ensemble to it. Not yet aware of Gibbs' work, he chose the microcanonical ensemble, as was usual. His justification was twofold: firstly, this type of ensemble is stationary[72]; secondly, except for an infinitesimal probability of error, "a system selected at random" from such an ensemble will be in a "normal state", *i.e.* that of thermal equilibrium and characterized by the Maxwell-Boltzmann distribution of the velocities.[73]

Since we cannot ascertain which system we actually select, or

"since the data of a problem of statistical mechanics do not completely specify the dynamical coordinates of the system, we are, in a problem of statistical mechanics, discussing an infinite number of different problems at once, and without differentiation *inter se*."

The *Stoßzahlansatz*, if projected onto this background, can then be seen to amount to the exclusion of an infinitesimal subclass of possible cases. The results obtained with its help are, therefore, only probable, though "infinitely probable".[74] In other words: the objective dynamical theory can be correctly used,

"provided it is interpreted with reference to the basis of probability supplied by our generalized space, and provided it is understood that it gives probable, and not certain, results".[75]

And the notions 'generalized space' and 'probable' are here to be understood as epistemic notions. This reinterpretation of the entire theory became a good deal clearer in Gibbs.

5.4. *Gibbs*

Two years earlier than Jeans, Gibbs had already introduced the ensemble approach as his basic tool from the very beginning; for him it is the adequate formal expression of a "statistical" inquiry[76] and the basis for defining measures of probability.[77] He barely notices earlier forms of statistical theory: (1) statistics of the particles of a system, (2) statistics of the phases of one system through time. The ergodic hypothesis is not even mentioned in this context. Instead Gibbs introduces, from the outset, an epistemological connection between the two theories that present our reduction problem:

"The laws of thermodynamics . . . express the laws of mechanics for such systems as they appear to beings who have not the fineness of perception to enable them to appreciate quantities of the order of magnitude of those which relate to single particles, and who cannot repeat their experiments often enough to obtain any but the most probable results."[78]

One might think that this idea is only meant to provide us with a background interpretation for the use of certain types of ensembles, whereas the proper justification of the latter is to be sought in the formal analogies of their behaviour with that of thermodynamical systems. In this case, no ontological reduction would even have been attempted, but its feasibility (*e.g.* by appealing to ergodicity) would be left open. Actually, though, Gibbs had to return explicitly to the epistemological basis of the ensemble approach: his proof of the H-theorem makes essential use of taking averages over *finite* cells in phase space, the justification being that, within phase cells "which are very small but not infinitely small", the density of systems in phase space "may in general be regarded as *sensibly constant*".[79] The indispensability of this assumption for the H-theorem as well as for finding statistical analogues of thermodynamical systems in general was stressed by the Ehrenfests[80]; and the justification in terms of limited observability has remained a standard argument in modern textbooks.

A second epistemological consideration is essential in Gibbs' derivation of irreversibility. He saw that this argument for the H-theorem applies to the past as well as to the future, so that he felt compelled to offer a further reason for restricting its application to the future. The temporal asymmetry, he suggested, is "immaterial with respect to mathematical fictions", but not "with respect to the events of the real world". Concerning the latter, "we are rarely justified in excluding the consideration of the antecedent probability of the prior events", *i.e.* our actual knowledge of the past.[81]

In one reading, at least, Gibbs appears to have given up the attempt at ontological reduction in favour of a novel and unified formal procedure, that of statistical physics in what was to become its standard form. In spite of its formal elegance and extreme fruitfulness, some physicists felt that it was lacking in the "sanity or physical reality" of Boltzmann's approach.[82] I take it that the incriminating short-coming could also be expressed by saying that Gibbs' approach failed to achieve an ontological reduction.

5.5. *Einstein*

It has been mentioned above (sec. 3, note 13) that Einstein did not share this rather sceptical view concerning the achievements of the kinetic theory of heat, though his approach is not essentially different from that of Gibbs. And I wish to conclude with a suggestion as to why this is the case. In the same way as Gibbs, he introduced the ensemble approach without much comment; he, too, selects the canonical (as opposed to the microcanonical) ensemble. But whereas Gibbs in the context of introducing it[83] recommended it only by the remark that it "seems to represent the most simple case conceivable", Einstein immediately connects it with a physical interpretation that is already contained in the title of the relevant section of his first paper[84]: "On the (stationary) probability of the states of a system S that is mechanically coupled with a system Σ of comparatively infinite energy". In other words: the idea of the heat bath provides an interpretation as well as physical motivation of the approach in terms of ensembles, indeed, exactly in terms of canonical ensembles. This reading is supported by the fact that Einstein connected his account with the explicit introduction of the thermometer as a particular kind of system *S* with the property quoted above.[85]

Neglecting all further details[86], one may say that Einstein in his early papers on thermodynamics anticipates his realistic way of shaping a theory in accordance with the *actual* situation of the experimenting physicist (cp. his analysis of simultaneity). It is the irony of this fruitful procedure that it harmonizes ill with Einstein's ontological convictions. Should it turn out that the truly physical intertheoretic relation between thermodynamics and (classical) particle mechanics demands the introduction of the heat bath in order to justify the ensemble, and thereby interpret the notion of probability in a non-subjectivist way, it would become impossible to dispense with notions like 'equilibrium' and 'temperature' as basic and irreducible ones. This alternative possibility of looking at the problem of relating mechanics and thermodynamics is also found in Gibbs; he explicitly says that it amounts to regarding "the temperature as an independent variable".[87]

6. CONCLUSION

The material presented in sections 4 and 5 may be briefly and preliminarily evaluated as follows: The early history of statistical mechanics stood under the *Leitmotiv* of reducing thermal phenomena entirely to mechanical theory. The problems of this programme arising from the (inevitable) introduction of probabilistic considerations were only belatedly and insufficiently recognized. They gave occasion to an overall trend in the changing views on probability: from more or less implicit quasi-aprioristic assumptions, through diverse physical and objective explanations on deterministic grounds, to the epistemic interpretation of ensembles.

None of these views turned out to provide a satisfactory foundation for an ontological reduction of thermal phenomena to motion or for the derivation of thermodynamics from mechanics without additional premisses which are either not generally true or not amenable to the interpretation required by the pursued ontological aim. A realistic conception of ensembles based on the idea of the "heat bath" can serve to eliminate the excess of epistemological elements from the interpretation of statistical mechanics, but it assigns a basic and irreducible status to some thermodynamical notions. The reduction, as far as it is eventually achieved within the period under consideration, then has to be understood not as a derivation, but as a different intertheoretic relation whose metatheoretical account presents a task for further study.[88]

Universität Bielefeld

NOTES

[1] The standard account is originally found in Nagel 1949 and 1961, Chap. 11. For a recent account that includes a number of refinements and generalisations cp. Schaffner 1977.

[2] Thus in the example to be studied here the problem was clearly conceived as an intertheoretic problem: Clausius was quite clear that thermodynamics had better be developed independently from mechanics (1857, and Engl. translation in Brush 1965, p. 112, cp. also Brush 1965, p. 24; Daub 1967, p. 294, and 1971).

[3] The term 'absorption' is Nagel's (1961, p. 337).

[4] Schaffner 1977.

[5] Darden/Maull 1977; for our present example cp. also Krüger 1976, where the "reduction" of thermodynamics to mechanics is interpreted as the development of a *third* theory, statistical mechanics. It is the purpose of the present paper to explore the historical background of this view somewhat more carefully and to give due attention to the key role of the notion of probability in this context.

[6] Maull 1977 attempts an evaluation of the unity of science in more general terms which is based on the idea of interfield theories. Cp. also Krüger 1980; and for the case of statistical mechanics Brush 1976 b.

[7] Nagel 1961, Chap. 11, part III. 3; Hempel 1969.

[8] Three recent publications are: Gillispie 1963, Garber 1972, and Schneider 1974/5.

[9] Hacking 1975 argues that a conceptual "duality" was an essential element in the very formation of our notion of probability in the 17th century.

[10] This last position has been adopted by Carnap, first 1945.

[11] Cp. esp. section 5 below.

[12] Zermelo was one of the few people who insisted in the incompatibility of reversible basic theory and irreversible macrotheory and, therefore, found himself compelled to point to the possibility of truly stochastic microprocesses (Zermelo 1896). Cp., also, Poincaré, who came to the conclusion "that the two principles of the increase of entropy and of the least action (in Hamilton's sense) are irreconcilable" and that for the explication of irreversibility one would have to "give up the familiar hypotheses of rational mechanics from which one has derived Lagrange's and Hamilton's equations" (Poincaré 1889, p. 233).

[13] Einstein 1902, p. 432.
[14] *ib.*, p. 417; the emphasis is mine.
[15] Garber 1972, p. 28, compares Boltzmann's underdeveloped awareness of the probabilistic nature of the second law with Maxwell's insight, who had invented his demon as early as in 1867, i.e. 8 years before Boltzmann for the first time noted exceptions to the second law (1875, p. 14). Cp. also Grant 1976, pp. 7, 9–10. A careful study of Boltzmann's repeated reinterpretations of probability is found in Klein 1973.
[16] Daub 1969, pp. 320 ff.
[17] Boltzmann 1877b, p. 164.
[18] Boltzmann 1871c, p. 295; emphasis added.
[19] Boltzmann 1872, pp. 316/7.
[20] Klein 1973, p. 62.
[21] Boltzmann 1868.
[22] Boltzmann 1871b, part III.
[23] Boltzmann 1872, p. 317.
[24] Boltzmann 1868, p. 50; cp. also *e.g.* pp. 64, 66, 81 for phrases like "the sum of the time moments" or "the time, hence the probability". There are many parallels in later papers, whereas I am not aware of phrases of the type "the number of molecules per volume, hence the probability": relative number is a self-explanatory concept.
[25] Boltzmann 1866, p. 13; 1868, p. 50; similarly p. 56; 1871a, p. 237; 1872, pp. 321–322.
[26] Boltzmann 1875, p. 2.
[27] *E.g.* Boltzmann 1866, p. 13; 1868, p. 50; and *passim* through many papers.
[28] The usual interpretation would have it that N is an *actual* number and only $\varphi(c)\,dc$ a probability. But what then is $f = N \cdot \varphi(c)\,dc$? f is certainly also a "number" very much like N. Moreover, this reading is at variance with Boltzmann's claim that he had always seen the probabilistic character of f (cp. notes 16–18).
[29] Good examples are 1871b, pp. 271–272, where the *same* mathematical expression is interpreted first as a "time [fraction]", and then as an "average distribution function" within a few lines. Cp. also 1872, p. 400, where he gets from a "number" f to a "probability", respectively "time ... divided by the total time" f/N by no more than a simple division by the total number N of molecules involved.
[30] Cp. especially the more striking cases mentioned in the last footnote.
[31] *E.g.* Boltzmann 1871b, p. 287.
[32] Boltzmann 1871b, part III; 1871c, pp. 293–294; 1872, pp. 401–402.
[33] Boltzmann 1872, p. 345, cp. also 1871a, p. 255. It is instructive to compare the latter passage with 1871b, p. 277, where he considers the possibility that the state of a gas might not be uniquely determined as a consequence of an ergodicitylike assumption.
[34] Boltzmann 1872, p. 317.
[35] The only statement in his earlier writings I am aware of that might be thought to suggest an epistemological background is "that we observe nothing but those average values" (1872, p. 316). But read in its context it appears to say that the treatment in terms of distribution functions is pragmatically sufficient, rather than that it could be justified in terms of our limited capacities of observation.
[36] That there could be several equilibria for given temperature and density is *a priori* unlikely – a gratuitous assumption – and not confirmed by any experience (Boltzmann 1871a, p. 255). Experience also shows that the phases of the internal motion of bodies that are brought into contact have no influence on their subsequent thermal behaviour

(1872, p. 401); the reason, according to Boltzmann, being "undoubtedly" that their effect is averaged out (cp. next footnote plus corresponding text).
[37] Translation from Brush 1976a, p. 366.
[38] Boltzmann 1871b, p. 284; 1872, p. 316. In the latter passage he points to the parallel with social statistics. Gillispie 1963 and Garber 1972 argue that this parallel was also involved in the history of Maxwell's ideas on probability. It is worth noting here that, also for Maxwell, the main reason for believing in the ergodic hypothesis was the existence of external disturbances of the motion of the ergodic system (see sect. 5.1. below).
[39] Klein 1973, sects. 5 and 7, who also points out that Loschmidt's objection as well as the need for a truly statistical reply to it had already been discussed in England several years earlier by William Thomson and Maxwell (sect. 6: cp. also Brush 1976, vol. 2, pp. 602 ff.); T. S. Kuhn 1978, Chap. II. I am greatly indebted to Professor Kuhn for letting me use this work prior to publication and – more generally – for introducing me into the study of the entire field.
[40] Boltzmann 1877a, p. 120.
[41] Boltzmann 1872, p. 317.
[42] Boltzmann 1877b, p. 164; emphasis is added.
[43] Klein 1973, p. 73.
[44] Cp. *e.g.* Boltzmann's own first example (1875, p. 14) and footnote 38.
[45] Boltzmann 1877b, p. 165.
[46] Brush 1976a, vol. 2, p. 610 quotes a phrase that might be thought to point in this direction. Furthermore, he reminds us, quite rightly, that the abandonment of dynamics is "the key to the power of the new method". In the (Boltzmannian) context of reduction it is rather a weakness.
[47] Köhler has pointed out the incompatibility of the two arguments for the H-theorem given in 1872 and 1877 (1931, pp. 344/5).
[48] Boltzmann 1896a, §6; the use of the method of 1877 in Boltzmann's later work remains marginal (1878, 1881a).
[49] Boltzmann 1877a, p. 119.
[50] Gibbs discusses this argument and resorts to epistemological remedies (1902, p. 151).
[51] Boltzmann did not easily accept the solution in terms of external disturbances as the general one (1895b); but he did not entirely drop it either: see 1898a, p. 633; 1898b, §91; cp. Brush 1976a, vol. 2, pp. 624–625, 636–637.
[52] Zermelo 1896.
[53] Brush 1976a, §14.7.
[54] Boltzmann 1895a.
[55] Boltzmann 1896b, p. 574.
[56] Cp. note 49 and related text.
[57] Boltzmann 1898b, §87; emphasis added. The translation is taken from Brush, p. 442.
[58] Boltzmann 1897, p. 579. This idea has been carefully worked out in the sequel: cp. Reichenbach 1956; Grünbaum 1963/73, Chap. 8. In particular, it is made clear by Grünbaum that the cosmological assumption is to be complemented by a further assumption that is, once again, of a *statistical* nature: the subsystems branching off from the wider universe in a given state of low entropy are *randomly* distributed over the microstates that are compatible with this entropy value (pp. 256f.).
[59] It is, perhaps, no more than a curiosity that Boltzmann calls the postulate

"assumption *A*"; it can indeed replace Burbury's "condition *A*" (but Boltzmann continues to use, alternatively, an assumption of this type, once again under the name "assumption *A*", Maxwell's reciprocity of collisions in the stationary state, 1898b, §92). The assumption concerning initial states is much more general, but also much closer to a petitio of the second law itself.

[60] Boltzmann 1895a; 1898a.

[61] Brush 1976a, vol. 2, p. 633.

[62] Ehrenfest 1910, §16.

[63] Boltzmann 1897, p. 584.

[64] Boltzmann 1898b, §89; emphasis added. I have quoted the German, because Brush's translation, though correct, does not reproduce the spatial metaphor of the 'side by side' of the two theories.

[65] Maxwell 1878, pp. 714/5.

[66] Boltzmann 1881b, p. 582.

[67] Rayleigh 1900, p. 111.

[68] See Brush 1976a, pp. 624–625. "Condition *A*" is essentially equivalent to the *Stoßzahlansatz* or to Boltzmann's requirement of "*molekulare Unordnung*".

[69] Jeans 1904; quoted after the *fourth* edition of 1925 which was the only one available to me. As far as the following remarks go, the differences between the diverse editions will probably not matter; a more careful study, though, will have to take them into account.

[70] *ib.*, Chaps. II–IV.

[71] *ib.*, §§125–128; the first edition contains this material in §§91ff.

[72] *ib.*, §39.

[73] *ib.*, §58. The phrase "selected at random" is a recurrent expression in Jeans.

[74] *ib.*, §138; cp. also §58.

[75] *ib.*, §70.

[76] Gibbs 1902, preface, pp. vii/viii.

[77] *ib.*, *e.g.* pp. 16–17.

[78] *ib.*, p. viii.

[79] *ib.*, p. 148; emphasis added.

[80] Ehrenfest 1906; 1910, sects. 12a, 23a.

[81] This passage was worked out in a strictly epistemological spirit by von Weizsäcker 1939; an alternative explication in terms of initial conditions *à la* Boltzmann (cp. part 4.3. of this paper) has been given by Grünbaum 1973, pp. 243 ff.

[82] Fowler 1936, p. 7.

[83] Gibbs 1902, p. 33; but compare p. 178!

[84] Einstein 1902, p. 420.

[85] *ib.*, §5; similarly 1903, §§3–4.

[86] These details call for a careful study that might of course modify the present rough interpretation considerably.

[87] Gibbs 1902, pp. 179–180.

[88] On the basis of his comprehensive studies that include, also, more recent material, *i.e.* the Ising model of phase transitions, Brush suggests a similar conclusion (1976b, esp. pp. 564–565).

REFERENCES

Boltzmann, Ludwig: 'Über die mechanische Bedeutung des Zweiten Hauptsatzes der Wärmetheorie', *Wien. Ber.* **53**, (1866); quoted from Boltzmann 1909/68, vol. I, 9–33.

Boltzmann, Ludwig: 'Studien über das Gleichgewicht der lebendigen Kraft zwischen bewegten materiellen Punkten', *Wien. Ber.* **58** (1868); quoted from Boltzmann 1909/68, vol. I, 49–96.

Boltzmann, Ludwig: 'Über das Wärmegleichgewicht zwischen mehratomigen Gasmolekülen', *Wien. Ber.* **63** (1871a; quoted from Boltzmann 1909/68, vol. I, pp. 237–258.

Boltzmann, Ludwig: 'Einige allgemeine Sätze über Wärmegleichgewicht', *Wien. Ber.* **63** (1871b); quoted after Boltzmann 1909/68, vol. I, pp. 259–287.

Boltzmann, Ludwig: 'Analytischer Beweis des zweiten Hauptsatzes der mechanischen Wärmetheorie aus den Sätzen über das Gleichgewicht der lebendigen Kraft', *Wien. Ber.* **63** (1871c), quoted from: Boltzmann 1909/68, vol. I, pp. 288–308.

Boltzmann, Ludwig: 'Weitere Studien über das Wärmegleichgewicht unter Gasmolekülen', *Wien. Ber.* **66** (1872); quoted from Boltzmann 1909/68, vol. I, pp. 316–402.

Boltzmann, Ludwig: 'Über das Wärmegleichgewicht von Gasen, auf welche äußere Kräfte wirken', *Wien. Ber.* **72** (1875); quoted from Boltzmann 1909/68, vol. II, pp. 1–30.

Boltzmann, Ludwig: 'Bemerkungen über einige Probleme der mechanischen Wärmetheorie', *Wien. Ber.* **75** (1877a); quoted from Boltzmann 1909/68, vol. II, pp. 112–148.

Boltzmann, Ludwig: 'Über die Beziehung zwischen dem zweiten Hauptsatz der mechanischen Wärmetheorie und der Wahrscheinlichkeitsrechnung respektive den Sätzen über das Wärmegleichgewicht', *Wien. Ber.* **76** (1877b); quoted after Boltzmann 1909/68, vol. II, pp. 164–223.

Boltzmann, Ludwig: 'Weitere Bemerkungen über einige Probleme der mechanischen Wärmetheorie', *Wien. Ber.* **78** (1878); quoted from Boltzmann 1909/68, vol. II, 250–288.

Boltzmann, Ludwig: 'Über einige das Wärmegleichgewicht betreffende Sätze', *Wien. Ber.* **84** (1881a); quoted from Boltzmann 1909/68, pp. 572–581.

Boltzmann, Ludwig: 'Referat über die Abhandlung von J. C. Maxwell "Über Boltzmanns Theorem betreffend die mittlere Verteilung der lebendigen Kraft in einem System materieller Punkte"', *Wied. Ann. Beib.* **5** (1881b); quoted from Boltzmann 1909/68, vol. II, pp. 582–595.

Boltzmann, Ludwig: 'On certain questions of the theory of gases', *Nature* **51** (1895a); quoted from Boltzmann 1909/68, vol. III, 535–544.

Boltzmann, Ludwig: 'On the minimum theorem in the theory of gases', *Nature* **52** (1895b); quoted from Boltzmann 1909/68, vol. III, p. 546.

Boltzmann, Ludwig: '*Vorlesungen über Gastheorie*, Leipzig, vol. 1 (1896a) and vol. 2 (1898b); eng. trans. by S. G. Brush, Berkeley: Univ. of California Press 1964.

Boltzmann, Ludwig: 'Entgegnung auf die wärmetheoretischen Betrachtungen des Hrn. E. Zermelo', *Wied. Ann.* **57** (1896b); quoted from Boltzmann 1909/68, vol. III, 567–578.

Boltzmann, Ludwig: 'Zu Hrn. Zermelos Abhandlung "Über die mechanische Erklärung irreversibler Vorgänge",' *Wied. Ann.* **60** (1897); quoted from Boltzmann 1909/68, vol. III, 579–686.

Boltzmann, Ludwig: 'Über die sogenannte H-Kurve', *Math. Ann.* **50** (1898a); quoted from Boltzmann 1909/68, vol. III, 629–637.

Boltzmann, Ludwig: *Wissenschaftliche Abhandlungen*, ed. Fritz Hasenöhrl, 3 vol., Leipzig 1909, repr. New York 1968.

Brush, Stephen G.: *Kinetic Theory*, vol. 1: *The Nature of Gases and of Heat*, Oxford 1965.

Brush, Stephen G: *The Kind of Motion We Call Heat – A History of the Kinetic Theory of Gases in the* 19*th Century*, 2 vols.; Amsterdam: North Holland 1976a. Most parts of the book have been published earlier as separate articles; for convenience I only quote their final version as contained in this work.

Brush, Stephen G.: 'Statistical Mechanics and the Philosophy of Science: Some Historical Notes,' *PSA 1976*, vol. 2, ed. F. Suppe and P. D. Asquith, East Lansing/Mich. 1976b, 551–584.

Carnap, Rudolf: 'The two concepts of probability', *Philosophy and Phenomenological Research* **5** (1945); repr. in: H. Feigl and M. Brodbeck (eds.), *Readings in the Philosophy of Science*, New York: Appleton-Century-Croft 1953.

Clausius, Rudolf: 'Über die Art der Bewegung, welche wir Wärme nennen', *Annalen der Physik* **100** (1857), 353–380.

Darden, Lindley, and Nancy Maull: 'Interfield theories', *Philosophy of Science* **44** (1977), 43–64.

Daub, Edward E.: 'Atomism and thermodynamics', *Isis* **58** (1967), 293–303.

Daub, Edward E.: 'Probability and thermodynamics – The reduction of the second law', *Isis*, **60** (1969) 318–330.

Daub, Edward E.: Rudolf Clausius; article in: *Dictionary of Scientific Biography*, vol. 3 (1971) 303–311.

Ehrenfest, Paul und Tatjana: 'Zur Theorie der Entropiezunahme in der statistischen Mechanik von Gibbs', *Wien. Berichte* [IIa] **115** (1906), pp. 89 ff.

Ehrenfest, Paul und Tatjana: *Begriffliche Grundlagen der statistischen Auffassung in der Mechanik*, 1910, Engl. Transl. Ithaca: Cornell Univ. Press 1959.

Einstein, Albert: 'Kinetische Theorie des Wärmegleichgewichtes und des zweiten Hauptsatzes der Thermodynamik', *Annalen der Physik* **9** (1902), 417–433.

Einstein, Albert: 'Eine Theorie der Grundlagen der Thermodynamik, *Annalen der Physik* **11** (1903) 170–187.

Fowler, R. H.: *Statistical Mechanics*, 2. ed., Cambridge Univ. Press 1936.

Garber, Elizabeth: 'Aspects of the introduction of probability into physics', *Centaurus* **17** (1972) 11–39.

Gibbs, J. Willard: 'Elementary principles in statistical mechanics', Yale UP 1902; quoted after the reprint: New York: Dover 1960.

Gillispie, C.C.: 'Intellectual factors in the background of analysis by probabilities', in: A. C. Crombie, *Scientific Change*, New York: Basic Books 1963, pp. 431–453.

Grant, C. T.: 'Boltzmanns statistische Interpretation des zweiten Hauptsatzes', *Humanismus und Technik* **20** (1976) 1–15.

Grünbaum, Adolf: *Philosophical problems of Space and time, New York* 1963, 2. ed. Dordrecht: Reidel 1973.

Hacking, Ian: *The Emergence of Probability*, London etc: Cambridge University Press 1975.

Hempel, Carl G.: 'Reduction: Linguistic and ontological facets', in: *Philosophy, Science, and Method: Essays in Honour of Ernest Nagel*, ed. *S. Morgenbesser et al.*, New York 1969, pp. 179–199.

Jeans, J. H.: *The dynamical theory of gases*, 1. ed., Cambridge 1904; quoted after the 4. ed. 1925.

Klein, Martin I.: 'The development of Boltzmann's statistical ideas', *Acta Physica Austriaca*, Suppl. **10** (1973): *The Boltzmann Equation*, ed. E. Cohen and W. Thirring, 53–106.

Köhler, W.: 'Zur Boltzmannschen Theorie des Zweiten Hauptsatzes', *Erkenntnis* **2** (1931) 336–353.

Krüger, Lorenz: 'Reduction versus elimination of theories', *Erkenntnis* **10** (1976) 295–309.

Krüger, Lorenz: 'Intertheoretic relations as a tool for the rational reconstruction of scientific development', *Studies in History and Philosophy of Science* **11** (1980), 89–101.

Kuhn, Thomas S.: *Black-Body Theory and the Quantum Discontinuity, 1894–1912*, Oxford U.P. 1978.

Maull, Nancy L.: Unifying science without reduction', *Studies in the History and Philosophy of Science* **8** (1977), 143–162.

Maxwell, James Clerk: 'On Boltzmann's theorem on the average distribution of energy in a system of material points', *Cambridge Philosophical Society's Transactions*, vol. 12, 1878; quoted after Maxwell 1890/1965, vol. II.

Maxwell, James Clerk: *The Scientific papers*, ed. W. D. Niven, Cambridge 1890, 2 vols., rep. New York 1965.

Nagel, Ernest: 'The meaning of reduction in the natural sciences' Chap. 4 of: *Science and Civilization*, ed. R. C. Stauffer, Madison/Wisconsin 1949, pp. 99–135.

Nagel, Ernest: *The Structure of Science*, London 1961.

Poincaré, Henri: 'Sur les tentatives d'explication mécanique des principes de la thermodynamique', *Comptes rendus de l'Académie des sciences* **108** (1889) 550–553, quoted after: Oeuvres vol. X, ed. G. Petiau, Paris 1954, pp. 231–233.

Rayleigh (John W. Strutt): 'The law of partition of kinetic energy', *Philos. Magazine* [5], **49** (1900) 98–118.

Reichenbach, Hans: *The Direction of Time*, Berkeley 1956.

Schaffner, Kenneth: 'Reduction, reductionism, values, and progress in the biomedical sciences', in: R. G. Colodny (ed.), *Logic, Laws, and Life – Some Philosophical Complications*, Univ. of Pittsburgh Press 1977, 143–171.

Schneider, Ivo: 'Rudolph Clausius' Beitrag zur Einführung wahrscheinlichkeitstheoretischer Methoden in die Physik der Gase nach 1856', *Archive for History of Exact Sciences* **14** (1974/5), 237–261.

von Weizsäcker, Carl Friedrich: 'Der zweite Hauptsatz und der Unterschied von Vergangenheit und Zukunft', *Ann. der Physik* **36** (1939), 275–283.

Zermelo, Ernst: 'Über einen Satz der Dynamik und die mechanische Wärmetheorie', *Annalen der Physik* **57** (1896), 485 ff.

ERWIN N. HIEBERT

BOLTZMANN'S CONCEPTION OF THEORY CONSTRUCTION: THE PROMOTION OF PLURALISM, PROVISIONALISM, AND PRAGMATIC REALISM*

Boltzmann's contributions to theoretical physics constitute a landmark in the history and philosophy of science. His papers and treatises that bear on statistical mechanics and the kinetic theory of gases occupy a position of seminal importance for the resolution of the crises in physics that characterize the development of early 20th-century scientific thought. The Boltzmann equation of 1872, namely the distribution function which gives the statistical probability of locating particles as a function of space, time, and velocity, is well known. Again, Boltzmann's fame is anchored to contemporary theory in physics by his clear and rigorous mathematical definition of the entropy function as related to probability and the statistical mechanics of atomic distribution. This is the celebrated H-theorem in which irreversibility is interpreted as an essentially statistical property of an ensemble of particles. In addition, we, of course, acknowledge the merits of Boltzmann's exhaustive exploration of the implications of the kinetic theory of gases, a subject which he pursued with determination and success for more than two decades.

The period of Boltzmann's greatest scientific productivity covers a 25 year span from about 1865–90. During this time he was, in turn, professor of experimental physics, mathematics, and then mathematical physics, first in Graz, then in Vienna, and again in Graz. During the last 15 years of his life, as professor of theoretical physics and the philosophy of science in Munich, Vienna, and Leipzig, Boltzmann turned his attention more and more toward philosophizing and deliberating about the object, methods, and limits of science. It is mainly about this philosophical and introspective aspect of his career that I would like to offer some comments. One way to search out the self-image of Boltzmann as a philosopher of science, and to discover the way in which he conceived of his own work and its relation to

J. Hintikka, D. Gruender, and E. Agazzi (eds.), Pisa Conference Proceedings, Vol. II, 175–198.

ideas within the context of the scientific currents of his day, is to listen to what he has to say as he reflects so persuasively on these matters in many of his essays and lectures.

Before we undertake to examine the ideas that stand out most conspicuously in Boltzmann's conception of theory construction, may I offer a few remarks about his general disposition toward the philosophy of science as a branch of learning. Like Mach, Boltzmann tells us that early in life he had an aversion and hatred for philosophy ("*ein Widerville, ja Hass gegen die Philosophie*"). This point of view, he believes, was shared by most natural philosophers (*Naturforscher*). Later in life, as he himself took to philosophizing about science, he wrote: "If it was only with hesitation that I followed the call to meddle with philosophy, philosophers have the more often meddled with natural science. They have now for many years invaded my preserves and I could not even understand what their views were and therefore wanted to improve my knowledge of the fundamental theories of all philosophy".[1]

Consistent with his distrust of any anxiously elaborated, inclusive, and systematic philosophy, Boltzmann – often with flippant self-reliance and articulate raciness – reveals, *ad libitum*, his deliberations, commitments, deep prejudices and conjectures about what, in his opinion, scientific activity is and ought to be. One gains the impression that the genesis of his contemplative views about science, or perhaps more correctly his own projected image of the scientist at work, is a reflective reconstruction, a philosophy of science, that is drawn mainly, and in retrospect, with the intent of justifying the way in which he has been doing science all his life. While this may not be an unbiased and objective account of the way that Boltzmann played the role of the scientist, it does represent a perspective that is worth examining. In fact, it is his self-image that provides the focus of our concerns here.

Boltzmann's appraisal of various philosophers is unmatched, to say the least. He offers the most derogatory comments about Kant, Hegel, Schopenhauer, and Herbart. The greatest positive influence upon his intellectual life he attributes to Schiller, Beethoven, Darwin, Goethe, and Shakespeare. Scientists who notably merit a prominent place in Boltzmann's philosophy of science include Galileo, Newton, Poisson, Laplace, Lagrange, Fourier, Hamilton, and Weber; but for unmitigated praise he singles out the work of Faraday, Mayer,

Clausius, Maxwell, Helmholtz, and Gibbs. The views of Hertz, Kirchoff, Mach, and Ostwald also are analyzed in considerable detail but far more critically.

In the face of these biting comments, Boltzmann damns philosophy with faint praise, and finally comes around to recognizing, as he puts it, that philosophy is the "greatest and true queen of the sciences"[2] with the help of which one may seek to resolve certain intellectually irresistible and perennially significant puzzles that are an intrinsic part of the whole scientific enterprise. But where is the illumination for this task to come from? How shall the philosophically untutored physicist (he means himself) acquire the necessary tools and information to engage in philosophy of science where the traditional philosophers have failed? In a Faustian mood he dismisses such academic questions by saying: "I am to teach with bitter sweat what I do not know myself".[3] Recognizing that his role has changed from that of a critic to an advocate of philosophy Boltzmann says in jest: "I should almost like to say that in my case the goat has been made the gardener".[4]

In spite of Boltzmann's relatively late excursion into the world of the philosophy of science, the development of theory unquestionably occupied the center of his life-long scientific interests. He says: "The idea that fills my thought and action . . . [is] the development of theory for whose glory no sacrifice is too great for me; since theory, is the whole content of my life".[5] If "theory" occupied so central a position in Boltzmann's thought and writings, then we must try to explore the meaning and status that "theory" takes within his philosophical world view.

What I aim to do in this paper is to furnish evidence to support the thesis that Boltzmann principally was challenged, in his scientific outlook, by the theoretical component of science, but that he took a rather dim view of achieving, within science, theories that are comprehensive, final, unitary, and unfalsifiable. Theories were not reflections of nature as nature really is. Boltzmann believed that the truths about nature are incredibly deep and difficult to fathom. He also felt that the prudent investigator might want to avoid tackling the most visible, and seemingly most important, scientific issues. Rather, more modestly, he would explore special problems that are manageable, and, above all, capable of being formulated with precision.[6]

Essentially Boltzmann suggests that, in moving toward some goal, the scientist would be wise to set forth on a path that is within reach rather than to choose the one that seems to represent the straightest logically conceivable solution. How nature behaves in the unknown is virtually unpredictable. To achieve a modicum of theoretical understanding of the nature of things by circuitous and zig-zag routes is not the exception but the rule. Small, genuine advances are sufficient; and frequently a detour can become a valuable point of leverage. He writes: "Thus natural science appears completely to lose from sight the large and general questions; but all the more splendid is the success when, groping in the thicket of special questions, we suddenly find a small opening that allows a hitherto undreamt of outlook on the whole".[7]

Quoting Schiller, Boltzmann writes that "only if each continues to work where and how one is able, can we come nearer to truth by busyness [*Beschäftigung*] that never flags, that slowly forms but never destroys, that to the building of eternities brings grains of sand, yet from the heavy debt of time wipes minutes, days and years".[8] Thus Boltzmann proclaims that the guiding star for "genuine theoreticians" – here he names Columbus, Robert Mayer and Faraday[9] – was to gain a picture of nature within the intellect by observant understanding, and uninterrupted, piecemeal, imaginative elaboration.

Boltzmann's emphasis, obviously, was on framing and solving special questions as precisely and rigorously as possible. For him there were no absolutely rock bottom solutions to explaining the behaviour of nature – no *nuda veritas* – in spite of the notable existence, as he said, of some sophisticated accomplishments in certain narrowly defined domains in physics. The triplet at the head of the Preface to the 1897 *Lectures on the Principles of Mechanics* captures Boltzmann's spirit in *toto.*

Bring' vor, was wahr ist;
Schreib' so, dass es klar ist
Und verficht's, bis es mit dir gar ist!

According to Boltzmann "all our ideas are purely subjective" ("*Alle unsere Vorstellungen sind rein subjektiv*"). While a theory is "a mental inner picture," ("*ein rein geistiges inneres Abbild*"), it may nevertheless attain to a high degree of abstract perfection.[10] In his characterization of the method chosen to formulate the principles of mechanics, Boltzmann, concurring with Hertz' point of view, writes:

> ... our thoughts are mere pictures of objects (or better signs for them), which at most have some sort of affinity [*Verwandtschaft*] with ... [objects] but never coincide with them but are related to them as letters to spoken sounds or written notes to musical sounds. Because of our limited intellects these pictures can never reflect more than a small part of objects.[11]

This being so, Boltzmann points out that the temptation to take the subjective mental pictures for what really exists, and for over-rating their permanence, is enormous. At its worst, Hegel represents a blatant example of belief in the reality of theories. Indeed, "It is in this sense that Hegel is said to have regretted that nature was unable to realize his philosophic system in its full perfection". The mathematician too, says Boltzmann, is blinded by the inner perfection of formulae. Written "with the blood of his heart", the "highest wisdom borders on supreme folly".[12]

These are harsh indictments against the reality of theory; but Boltzmann is convinced that the inevitable dangers inherent in the lure of completeness of theories and the tendency to overshoot the mark in our enthusiasm for equating theories with reality puts the theoretical physicist in a particularly vulnerable position. He says: "If at the outset I have declared myself an advocate of theory, I will not deny that I have experienced the evil consequences of its spell".[13] For Boltzmann, theories alone can help us to explain nature, but they can explain only a few incredibly restricted domains of an immensely complex world.

In 1905 in his address at the St. Louis exhibition, Boltzmann mentioned that the most recent developments in experimental physics had overshadowed advances in theoretical physics. As he saw it, new phenomena had been revealed whose theoretical understanding was becoming increasingly obscure. In fact Boltzmann asserts in this context that the task of constructing lasting theories is almost hopeless. He comments that indefinite, mathematically shrouded theories are more secure but also more sterile than bold theories. He writes:

> It here becomes evident that in a certain sense experiment is entitled to precedence over all theory .... The conflict of theories, however, is an infinitely lengthy business, indeed it almost seems as though certain controversies as old as science itself will live on as long as it does.[14]

As for identifying Boltzmann's philosophical position on the side of the idealism-realism debate, we see that his position is not an unequivocal one. I think we can say, for example, that Boltzmann, contrary to Einstein, Planck, Schrödinger, and Heisenberg, was thoroughly convinced that while unitary, realistic, comprehensive, and over-arching theories of completeness in science were seductive, they often were elusive and sterile. I see that there are some striking similarities between Boltzmann and Mach on the status of theories in science; but there are, as well, crucial differences that I will try to clarify later on. Perhaps Boltzmann was closer to Bohr's position – who incidentally, like Boltzmann, often quoted aphorisms from Schiller; *exempli gratia*:

> *Nur Beharrung führt zum Ziel,*
> *Nur die Fülle führt zur Klarheit*
> *Und im Abgrund wohnt die Wahrheit.*[15]

In his attempt to come to grips with the rising tide of critical positivism, Lenin, after the abortive revolution of 1905, in his philosophical analysis, *Materialism and Empirio-Criticism*, counted Boltzmann among the opponents of those scientists whose philosophical views were leading the Marxist Machists astray. Lenin was disappointed to see that "Boltzmann . . . was afraid to call himself a materialist and even explicitly stated that he did not deny the existence of God". Nevertheless, Lenin was convinced that, like the majority of 19th-century scientists, Boltzmann, "whose theory of knowledge is essentially materialistic", mistrusted, and "systematically combated" the new epistemological dogmas of phenomenalist physics and the philosophical subjective idealism of "Mach and Co."[16]

Now I would suggest, *pace* Lenin, that Boltzmann was indeed a philosophical idealist on the subject of theory construction; but that he was a pragmatic realist when he came to consider the approach which scientists should adopt in their work that would be most conducive to high-level theoretical accomplishment. Boltzmann seems to have believed, or at least to have held *philosophically*, that the notion of objective existence was rather untenable and, at minimum, inaccessible. Accordingly, there were only two categories of existence to be established, *viz.* the subjective existence of mental images resulting from sensations, and non-existence *per se.* Nevertheless, Boltzmann was quite convinced, that for pragmatic reasons, one needed to operate as well with the categories of objective existence and objective non-existence.[17] For example, he tells us that he once was on the verge of the "tempting mistaken inference" of an unreal world.

Solipsism is the view that the world is not real, but a mere product of our fantasy, like a dream object. I too once hankered after this whim, which led to my failing to take the right practical action and caused me great damage, to my immense delight, for this provided me with the desired proof that the external world exists, a proof that can consist only in showing that if we doubt this existence we are less able to act appropriately.[18]

Where could you find a more striking confession of belief in pragmatic realism?

Labels aside, Boltzmann entertained the plausible notion that

the intimate connection of the mental with the physical is in the end given to us by experience,... [and that] by means of this connection it is very likely that to every mental process there corresponds a physical process in the brain, that is, there is an unambiguous correlation; and that the brain processes are genuinely material, that is, are representable by the same pictures and laws as processes in inanimate nature.... All these circumstances make it extremely likely that an (objective) world picture is possible.... [However] our aim will not be to establish the truth or falsehood of one or the other world picture, but we shall ask whether either is appropriate for this or that purpose while we allow both pictures to continue alongside each other.[19]

In spite of Boltzmann's play on objectivity, I call this pluralism.

Having suggested that one may proceed synthetically to generate a physical world picture from subjective experience – that is, one that connects the mental to the physical – Boltzmann entertains, as well, the reverse argument: *viz.* the shift from a subjective introspective construction of a world picture to a physico-physiological point of departure. This corresponds to the laying bare of concepts as precisely as possible, to show "that mental processes are identical with certain material processes in the brain". Boltzmann calls this "realism", and adds "It has often been held that this is impossible. Whether we are entitled to hold this view we can naturally test once again only from what is given in experience".[20] In this analysis, the psycho-physical parallelism embraced by Boltzmann is virtually identical with Mach's position *vis-à-vis* the mental and the physical.

It should be mentioned in this context that Boltzmann's recognition of the incredible richness of sensations is all of one piece with his belief in the limitless complexity of natural phenomena. He says:

It is precisely the protean variability in the nature of different sensations and their almost indefinable similarities that make it likely that their course cannot be represented by the most simple mental pictures but only by very complicated ones, like the various physical and chemical process in the brain.[21]

I would like at this point to discuss briefly Boltzmann's pluralism in relation to theories. Boltzmann, we have mentioned, had an extremely high opinion of the role of theory in science. In order to be most fruitful, he felt that theories should be formulated, initially, to explain phenomena for limited, special domains. They should move in the direction of comprehensiveness over wider domains only with caution and against the test of experimental verification. Thus the theories should not be too ambitious, should never be given the stamp either of finality or reality, and should not be so abstract, as to gloss over that deep complexity of nature which lies beyond all of the positive achievements of science at any one time.

It is unmistakably clear that Boltzmann was not one to search for overall theoretical unity in science. As for constructing world pictures, he was essentially a pluralist. He says:

It is still possible to have theories that correctly represent a large number of facts but are incorrect in other aspects so that they may have a certain relative truth. Indeed, it is even possible that we can construct a system of pictures of experience in several ways. These systems are not equally simple, nor represent phenomena equally well.... However, it may be doubtful and in a sense a matter of taste which representation satisfies us most. By this circumstance science loses its stamps of uniformity.

Error is manifold, but so is truth. Boltzmann continues: "According to the one view we turn our gaze more towards the one unattainable unitary theory, according to the other towards the multiplicity of what is attainable."[22] Boltzmann, assuredly, threw his lot in with the latter.

A good example of Boltzmann's tolerance for the simultaneous attack on scientific problems from different points of view is to be gleaned from his several articles on atomism. It is well known that Boltzmann was a great champion for atomism in the natural sciences.[23] What is not so commonly stressed is that he was well informed about and willing to entertain the value of exploring the various competing continuum theories that were advocated in his day. Thus he discussed at some length both the merits and inadequacies of the two current phenomenological alternatives to atomism – the mathematico-physical continuum approach, and the particular brand of thermodynamics promoted by the energeticists. He simply was anxious to keep all viable theories afloat, recognizing that a world picture with Fourier-like elegance eventually might be achieved phenomenologically or atomistically.[24] He says elsewhere: "The question whether matter consists of atoms or is continuous reduces to the much clearer one, whether the continuum is able to furnish a better picture of phenomena."[25]

Thus Boltzmann felt that

> nothing will be lost in certainty, while perspicuity will be retained, if we strictly separate the phenomenology of the best-attested results from atomistic hypotheses that serve comprehensiveness, both being further developed with equal vigour as being equally indispensable, rather than assert with one-sided regard for the advantages of phenomenology that one day it will certainly displace current atomism.[26]

This last remark about the indispensability of atomism in natural science, made in 1897, represents one component in the development of Boltzmann's defensive position in the midst of increasing attacks on the atomic and the kinetic-molecular theory of matter. In the Preface to the second part of his *Vorlesungen über Gastheorie* of 1898 he wrote:

> I am convinced that these attacks rest solely on misunderstandings, and that the role of gas theory in science is far from having been played out.... Therefore according to my opinion it would be a tragedy for science if the theory of gases were to fall into oblivion because of the currently fashionable hostile attitude toward it.... I am conscious of how helpless the individual is in struggling against the stream of time. But I have written this book... in order to contribute, what lies in my power, in such a way, that when the theory of gases is revived not too much will have to be rediscovered.[27]

Boltzmann, we know, compared and weighted the competing phenomenological theories with atomism, and maintained consistently that only the criterion of fruitfulness should be invoked to evaluate the various alternative theories. He felt that the atomic theory had achieved a number of positive accomplishments (in truth they were not overwhelmingly impressive) and that phenomenology – although a conceivable and explore-worthy alternative – had been fostered by investigators who had avoided dealing with special questions that could be formulated with precision. His criticism was that the phenomenologists had sought out, instead, general and hopelessly complex problems with the objective of furnishing all-encompassing pictures of nature without bothering about the details. Above all, Boltzmann maintained that the phenomenologists had drawn rather meager consequences from their theories; that they were engaged in spinning out irresponsible overly-ambitious systematic schemes. They were engaged in metaphysics, that is, – a relatively naughty word in Boltzmann's vocabulary. In fact, they were prone to raise their theories to the status of dogmas and had done so without having provided

an unambiguous, and comprehensive description of a single domain of stationary phenomena by means of this method, let alone a survey of phenomena including turbulent ones. The question whether this path will one day lead to comprehensive pictures of nature is thus for the time being only academic.[28]

Boltzmann's predilection was to advocate a range of special atomisms in which he attributed to the atoms in question only such properties as were needed to describe small factual domains as simply as possible and, therefore, reasonably free from arbitrary features. It is absolutely essential to recognize that for Boltzmann the atom was an extremely variable-prone, *i.e.*, flexible idea. He evidently was prepared to allow all necessary modifications, and even seemingly arbitrary features, to become a part of the theory. As for our representation of atoms he wrote:

It is precisely the main task of science to fashion the pictures that serve to represent a range of facts in such a way that we can predict from them the course of other similar facts .... There is then some hope that one might modify and perfect pictures in such a way that they do justice to the new facts too. (We learn something new about the constitution of atoms).[29]

In another connection, in his analysis of the "internal picture or mental construction", ("*ein inneres Bild, eine gedankliche Konstruktion*") that had been enunciated for the theorems of mechanics, Boltzmann voiced the same sentiment: "The ways in which we reached the pictures are not infrequently most disparate, and contingent on the most varied accidents." Boltzmann, evidently wanted to keep all useful options open and accessible; for he says: "in the end the only judge as to whether a theory is usuable remains experience, and its evident verdict cannot be shaken or appealed against."[30]

The particular features of natural phenomena that can be subjected to a mechanico-geometrico, *i.e.*, spatio-temporal analysis, is an open and challenging question for Boltzmann. Thus he admits that thermal phenomena are not so easily absorbed into the atomistic world-view; and even more doubt exists as regards electro-magnetic phenomena, radiant energy, and chemistry:

Thus even purely physical facts are by no means all of the same kind.... Mental phenomena [*psychische Erscheinungen*] may well be more remote from material ones than thermal or electrical from purely mechanical ones, but to say that the two former are qualitatively while the latter three are only quantitatively different seems to me mere prejudice.... If one thus adheres to this... view... processes in inanimate nature differ so little in quality from animate ones, that it is impossible to draw any boundary and it would be impracticable to ascribe objective existence only to sensations but not to processes in inanimate nature.[31]

In the same context Boltzmann implies that a content-rich concept of the atom, or of God for that matter – irrespective of the question of their existence – is worthy of being upheld, provided that one's goal is to achieve a world picture that attempts to encompass as many phenomena of nature as possible. I suggest that Boltzmann and Mach held views of atomism, and of God, that were essentially compatible. The difference in their views was that Boltzmann felt that both of these concepts were fruitful, or at least potentially fruitful of worthwhile consequences, whereas Mach did not. For both, we must note, 'atom' and 'God' were *concepts*, and, the question concerning objective existence was either unessential or at most a heuristic device to get on with the job of comprehending the nature of things.[32]

According to Boltzmann the phenomenologists were just barking up the wrong tree. "Phenomenology", he says,

tries to combine these special atomisms without prior simplification, in order to represent actual facts, that is in order to adapt all conceptions contained in these atomisms to the facts; however, since these are countless concepts severally taken from small factual domains and hardly compatible, along with countless differential equations each with its own peculiarities in spite of many analogies, we must from the outset expect that the representation will turn out very complicated. Indeed we find that if phenomenology is to do no more than represent the interlinkage between a few domains of phenomena of quasi-static processes (elastic deformation with heating and magnetisation and the like) one needs already very unwieldy and enormously complicated equations. Besides, even then one has to introduce hypotheses and thus go beyond the facts (for example if one wants to represent the dissociation of gases after Gibbs and that of electrolytes after Planck).[33]

Now Boltzmann was not opposed, in principle, to the search for comprehensive theories. Indeed, he recognized Fourier's theory of heat conduction as ideal in this regard. In order to get closer to such comprehensive pictures of nature, Boltzmann mentioned, with approval, that the atomists had sought to reconcile the foundations of the various phenomenological atomisms by arbitrarily completing and altering the properties of the atoms required for the various factual domains in such a way that they may serve for representing many domains at the same time.[34]

In effect, Boltzmann saw that the phenomenologists owed their professed success to the atomistic origins that had been stripped away in subsequent formulations. To be specific, atomism, he claimed, had been an indispensible crutch for the mathematics of heat conduction, elasticity, and electromagnetism. Notwithstanding the adequacy of treatment of some general issues, he believed that this phenomenological approach inevitably reached dead ends whenever the details were pursued – additional features thus having to be superimposed on the phenomenology in order to achieve any success at all.

In Boltzmann's reflections about atomism we are confronted with the timeless fascination connected with the role and characteristic province and inter-relationship of macro- and micro-conceptions of natural phenomena. From the side of phenomenological energetics, for example, the claim was being made in Boltzmann's day that classical thermodynamics provided all that was necessary to establish the criteria for spontaneity and equilibrium for chemical reactions – and, namely, without any reference to atomism, the micro-nature of the interacting substances, or the mechanism (*i.e.* kinetics) of the reactions. Nevertheless, Boltzmann's kinetic-atomistic approach, and his imaginative and rigorous formulation of statistical mechanics eventually silenced his anti-atomistic critics. By providing distribution functions which give the statistical probability of locating atoms and molecules at certain space-time positions and velocities, he led the way to establishing a statistical mechanics that, in time, encompassed the domains of both chemical reactivity and chemical kinetics.

We may note in connection with this discussion of Boltzmann's thoroughgoing reformulations and innovative moves within classical physics, that he entertained no inflated view of a grandiose scientific revolution. Rather, it was far more consistent with Boltzmann's world view to bring to light the multiplicity of revolutionary permutations that were taking place within the whole spectrum of the physical sciences. It was within this context that he encouraged the exploitation, to the limit, of conceptual schemes and models that had proved their worth in the deep exploration of natural phenomena. I may suggest that Boltzmann's creative approach to physics provides us with an example of an investigator who – not hung-up on specific themes or paradigms – was able effectively to correlate and integrate, and manipulate ideas that seemingly belonged to different compartments of thought.

According to his own autobiographical remarks, Boltzmann was an old-time, classical, 19th-century physicist working in the midst of other investigators whose activities, by contrast, revolved about all the many changes in science that had taken place at the turn of the century. Thus in 1899 he wrote:

> Indeed, when I look back on all these developments and revolutions [*Umwälzungen*] I feel like a monument of ancient scientific memories. I would go further and say that I am the only one left who still grasped the old doctrines with unreserved enthusiasm [*mit voller Seele*] – at any rate I am the only one who still fights for them as he can. I regard as my life's task to help to ensure, by as clear and logically ordered an elaboration as I can give of the results of classical theory, that the great portion of valuable and permanently usuable material that in my view is contained in it need not be rediscovered one day, which would not be the first time that such an event had happened in science.

Reminiscent of the self-justificatory statement that preceded the second part of the *Vorlesungen über Gastheorie*, quoted above, Boltzmann writing here, just one year later, says:

> I therefore present myself to you as a reactionary [*Reaktionär*], one who has stayed behind and remains enthusiastic [*schwärmt*] for the old classical doctrines as against the men of today; but I do not believe that I am narrow-minded [*borniert*] or blind to the advantages of the new doctrines, which will receive due justice in the next section of my talk so far as lies within my power; for I am well aware that like everyone else, I see things subjectively tinged through my own spectacles [*meine subjektiv gefärbte Brille*].[35]

We see that Boltzmann presents himself as one who had spent a lifetime in the imaginative, even romantic cultivation and elaboration of the consequences and implications of the major theoretical advances of 19th-century physics. Above all, he has been puzzling about the way in which specific methodological questions had illuminated theoretical physics. As for the future, he tells us that what he will have to say will be negative. In retrospect, at least, these deliberations strike us as strange self-image projections, especially coming from one who effected so monumental a transformation in the theoretical perspective of physics at the turn of the century.

The great conceptual scientific landmarks that Boltzmann singles out for serious evaluation (call them scientific revolutions if you like; Boltzmann's term is '*Umwälzungen*') are five in number. He first mentions Weber's electromagnetic particle-oriented theory, and places it alongside its elegant competitor, Maxwell's field theory. How completely characteristic of the development of the most portentous of theories, concludes Boltzmann, that

Maxwell wished his theory to be regarded as a mere picture of nature, a mechanical analog as he puts it, which at the present moment allows one to give the most uniform and comprehensive account of the totality of phenomena.

Boltzmann is quick to point out – true to his predilection for the fertility of a pluralism of theoretical frameworks – that Weber and Maxwell demonstrated "how a group of phenomena can be explained in two totally different ways, both modes of explanation representing the facts equally well".[36]

In second place Boltzmann recognizes the potential theoretical significance of the celebrated discovery of Röntgen rays. For if it really could be demonstrated that they were incapable of polarization, interference, and refraction – as seemed to be the case – then these "rays" either would have to be longitudinal ("if they be waves at all") or conceivably particulate in nature. Boltzmann suggests here that the solution to this enigma may help to clarify, as well, the transmission of longitudinal "waves" in the luminous aether. This picture, he admits, may not survive but "even an incorrect theory can be useful, if only it contains leads to novel experiments [*Anregungen zu neuen Versuchen*]".[37] What Boltzmann meant when he contrasted "correct theories" with "incorrect but useful theories" in this context is ostensibly clear. He says:

No theory can be objective, actually coinciding with nature, but rather . . . each theory is only a mental picture of phenomena, related to them as sign is to *designatum*. From this it follows that it cannot be our task to find an absolutely correct theory, but rather a picture that is as simple as possible and that represents phenomena as accurately as possible.[38]

As a third major landmark in classical physics, Boltzmann mentions Kirchoff's elimination of the concept of force, thus providing an alternative to the use of obscure concepts (*dunkle Begriffe*) and metaphysical considerations (*metaphysische Betrachtungen*). Hertz' mechanics, too, is mentioned as an innovative theory that obviously does not coincide objectively with nature, but is structured essentially on mental pictures. Boltzmann believed that it was conceivable that "Hertz' mechanics is more a programme for the distant future", because he felt that the theory was then mainly of academic interest, and had not yet demonstrated its problem-solving power.[39]

Another supremely important domain which Boltzmann set apart by virtue of its great wealth of brilliant results – *viz.* such as had "led to an attempted revolution involving the whole of physics" – pertains to the principles of energy and special (*i.e.*, classical) thermodynamics. More specifically, Boltzmann had in mind the development of general (*i.e.*, molecular) thermodynamics. This was associated with the entropy function notably elaborated by Clausius and Gibbs, and had been applied successfully to such diverse fields as capillarity, chemistry, and electric, magnetic, and radiant phenomena. The perturbing aspect of these developments was that they "caused such enthusiasm amongst some scientists who called themselves energeticists, that they taught that we must break with all past conceptions....", suggesting in particular that all mechanical models were superfluous and even injurious and that the whole of future natural science was a mere description of the behavior of energy. Now, Boltzmann, without renouncing the plausibility and pertinence of the energetic position, maintained that no matter how central the energy concept turned out to be, scientists should be wary of tolerating deceptive and superficial formal analogies from which inferences were drawn that would be less precise and more ambiguous than those that are drawn from classical physics.[40]

Finally, Boltzmann took the chance – in his 1899 paper on developments in recent theoretical physics – to offer some critical comments about the main stream of the "moderate secessionists", (*gemässigte Sezessionisten*) the phenomenologists, who seemed to him to be so much in ascendency in his day. I will just mention briefly in this connection, that Boltzmann's targets here were mainly Hertz' mathematical phenomenology – the most extreme form of phenomenology he calls it – and Mach's general phenomenology. As for the former, Boltzmann stresses both the impotent heuristics of pictureless mathematics, and the scientific sterility of writing down equations without derivation, comparing them with phenomena, and then regarding agreement between the two as the best proof that the equations are right, As for Mach, in his general phenomenological approach, he sought:

to describe every group of facts by enumeration and by an account of the natural history of all phenomena that belong to that area, without restriction as to the means employed except to renounce any uniform conception of nature, or any mechanical explanation on other rational foundation.

The heart of Boltzmann's criticism of the approaches of both Hertz and Mach was that they believed that phenomenology "could represent nature without in any way going beyond experience".[41]

Boltzmann attacks the phenomenologists for trying to link up with experience at the outset of an investigation rather than starting with hypothetical mental pictures. He says:

No equation represents any process with absolute accuracy, but always idealizes them, emphasizing common features and neglecting what is different and thus going beyond experience. That this [*i.e.*, the theoretical picturing or idealization] is necessary if we are to have any ideas that allow us to predict something in the future, follows from the nature of the intellectual process itself, consisting as it does in adding something to experience and creating a mental picture that is not experience and therefore can represent many experiences.

As Goethe says, he continues,

The more boldly one goes beyond experience, the more general the overview one can win, the more surprising the facts one can discover, but the more too one can fall into error. Phenomenology therefore ought not to boast [*prahlen*] that it does not go beyond experience, but merely to warn against doing so to excess.[42]

Boltzmann thought that Hertz and Mach were entertaining a far too grandiose opinion of the scientist's ability to go directly from the given phenomena (sensations, experiences) to the potential nature of things in the unknown; namely by attempting to do so without the use of models, pictures, and theories that would help to bridge the gap in going from the known to the unknown. To put it bluntly, they were blind to understanding that the real guts of science lies not just in playing with the phenomena in a clever, classificatory, abstract, or mathematical way, but rather in exerting the most strenuous and imaginative efforts to the framing of models, internal mental pictures, and theories that guide the investigator to a correct represenatation of phenomena – thus helping him to win his way through to those novel perspectives that go beyond the facts of experience, and that are inaccessible by any other route.[43] Concerning the status of statistical mechanics, for example, Boltzmann writes:

This theory does indeed boldly transcend experience but it has precisely the feature that such theories should have, namely to show the facts of experience in quite a new light and thus to spur one on to further reflections and research.[44]

A good example of the myopic perspective of the phenomenologists, in Boltzmann's opinion, was their inability to recognize how many new insights they failed to grasp with their special energetic thermodynamics by excluding atomic-molecular conceptions. To be specific, how could they fail to identify the achievements of atomism in relation to van der Waals' work on the equations of state, the ratio of heat capacities at constant pressure and volume in connection with the discovery of monatomic argon, Gibbs' theory of dissociation, the equations of hydrodynamics and thermal conduction, the entropy theorem in relation to the calculus of probability, and many other facts associated with physics, chemistry, and crystallography? Never blind to the potential shifting nature of scientific models and theories, Boltzmann remarks: "As regards ideas about the nature of molecules it will however be necessary to leave the widest possible room for *manoeuvre*". The paradox of this situation, as Boltzmann conceives it so cogently, is

that almost all equations that to the phenomenologists now count as the very essence of physics had been obtained by way of molecular theory, but that it had become superfluous now that we already have these equations.[45].

However, historical principles are sometimes double-edged. "Certainly", says Boltzmann,

history often shows unforeseen revolutions [*unvorhergesehene Umwälzungen*]; certainly it is useful to keep in mind the possibility that what now seems to be most secure may one day be displaced by something quite different; but equally the possibility [exists] that certain achievements will remain permanently established in science even if in amplified and altered form.[46]

As Boltzmann would have it, we do not know how permanent, how simple, and how comprehensive our scientific pictures can and will be in the future. For Boltzmann these are all fascinating questions, but we must not, to use his favourite expression "overshoot the mark" (*über das Ziel hinausschiessen*), for "it is better to work on what lies close to hand rather than rack one's brain about such remote questions".[47]

Remote questions, for Boltzmann, we must note, were not just questions for which there is a meagre empirical base, they were questions for which no precise, manageable, or operationally conceivable method of solution had been formulated. What I mean to emphasize in this context is that Boltzmann was no crass empiricist on the subject of theory construction; rather, he championed, at various times, for certain fields, what he called the deductive construction of internal mental pictures – "a particular mode of representation that has quite peculiar advantages, though it has defects too". Boltzmann gives us an explicit *exposé* of this deductive view of representation:

This mode consists in starting to operate only with mental abstractions .... In this we do not yet take account of facts of experience. We merely endeavour to develop our mental pictures as clearly as possible and to draw from them all possible consequences. Only later, after complete exposition of the picture, do we test its agreement with the facts of experience; it is, then, only after the event that we give reasons why the picture had to be chosen thus and not otherwise, a matter on which we give not the slightest prior hint. Let us call this deductive representation. Its advantages are obvious. For a start, it forestalls any doubt that it aims at furnishing not things in themselves but only an internal mental picture, its endeavours being confined to fashioning this picture into an apt designation of phenomena.[48]

Perfect clarity and absolutely rigorous consistency are Boltzmann's passwords. He says:

For it is one of the most important requirements that the pictures be perfectly clear, that we should never be at a loss how to fashion them in any given case and that the results should always be derivable in an unambiguous and indubitable manner. It is precisely this clarity that suffers if we bring in experience too early [*diese Klarheit leidet durch zu frühe Vermischung mit der Erfahrung*], and it is best preserved if we use the deductive mode of representation. On the other hand, this method highlights the arbitrary nature of the pictures since we start with quite arbitrary mental constructions [*die Willkürlichkeit der Bilder*] whose necessity is not given in advance but justified afterwards. There is not the slightest proof that one might not excogitate other pictures [*willkürliche Gedankenkonstruktionen*] equally congruent with experience.[49]

Both Maxwell and Hertz approximate to Boltzmann's image of this deductive mode of theory construction; but whereas Maxwell is praised for his magnificient accomplishments, Hertz' efforts are reviewed with cool reserve. Essentially the critique of Hertz' attempt was that it turned out to be too cumbersome to go from his deductive picture to a true and useful representation of fairly common phenomena. When Boltzmann was criticized by Johannes W. Classen for carrying out a polemic against Hertz and his views on mechanics – within the context of Boltzmann's position as expressed in his 1897 *Lectures on the Principles of Mechanics* – Boltzmann responded:

Nothing could be further from the mark. I absolutely recognize the advantages of Hertz' picture, but on the principle that it is possible and desirable to set up several pictures for one and the same group of phenomena, I think that alongside Hertz' picture mine still has its significance, in that it has certain advantages lacking in his.[50]

Boltzmann attempted to clarify his position by saying that Hertz' picture was superior in that "it contains many more features that are marked by an absence of inner necessity", whereas his own picture – which he characterized as "not essentially different from the other theories of mechanics" – was superior in representing "the most ordinary processes of daily experience involving the action of forces" – like gravitation, and the quite trivial case of elastic impact between two solid spheres. The real problem with Hertz' mechanics, in spite of its deductive simplicity, as Boltzmann saw it, was that one would have to develop a completely new theory to account for such simple phenomena as gravitation and elastic impact.[51]

The message that comes through with considerable transparency in Boltzmann's discussion of the fundamental principles and equations of mechanics, is his acknowledgement of a deep and consistent pluralism in regard to theory construction. He says:

For the time being we are not aiming at a single best account of science, but ... we regard it as expedient to try as many approaches as possible, each of which has its peculiar advantages but also its drawbacks.

In substance Boltzmann encourages the scientific investigator to try anything that is plausible as hypothesis formulation, but to do so with precision, logical rigour, and internal consistency. Above all no tacit concepts and assumptions should be smuggled in. The scientist constantly must be on the alert in regard to the specific hypotheses he fabricates and comes to rely on, otherwise falsifiability is difficult if not impossible.[52]

Boltzmann was firmly convinced that atomism was experientially more plausible and philosophically more satisfying than the continuum theory of matter. If I may offer a comparison between Boltzmann and Mach on this score I would put it as follows: Mach read the world and its deep complexity correctly, but seized the wrong end of the methodological argument to explain why he read the world as he did. In fact, he entertained an abominably lean opinion of the role of theory construction in science. If you start with the facts of experience, as Mach did, you necessarily are led to fabricate an integrated picture of natural phenomena that is irresponsibly narrow in its potential pertinence for any aspects of nature not included in the original facts.

As Boltzmann puts it in one place:

When Goethe says that only one half of our experience is experience, what he intends to convey by this seemingly paradoxical dictum is surely that in every conceptual grasp of experience or verbal representation of it we must already go beyond experience.[53]

In contrast to Mach's empirical and myopic point of departure, Boltzmann begins with mental constructs. In that case, one is in possession of a built-in manageable programme that can be manipulated as one seeks to check the nicety and precision of the fit of the theory with new facts. In truth, a phenomenological and inductively constructed sophisticated mechanics, for example, is a practical impossibility.

Unlike Mach, Boltzmann believed firmly in the ultimate indispensability and significance of mechanics as a basis of natural science and scientific theories as a whole. He says:

If therefore on the one hand it has indeed so far been impossible to prove that all natural science can be mechanically explained, on the other it is equally certain that nobody has shown certain phenomena to be not thus explicable; at most we can say that for certain phenomena an attempt at mechanical explanation is as yet premature.[54]

Let us remember, in light of Boltzmann's own fundamental contributions to statistical mechanics, that his conception of mechanics was very large and open indeed; and that in a very real sense his views on mechanics did provide the gateway to modern theoretical physics – at least as it is now constituted.

Superimposed on Boltzmann's resolute emphasis on the centrality of mechanics in physics was his observation that field theoretic views were providing a conceivable alternative world view. He says in 1900:

While ... the principle of mechanical explanation has conquered ever wider domains within the whole sphere of science, oddly enough it has lost some ground in its central field, namely in theoretical physics.[55]

What Boltzmann had in mind was the way in which the hegemony of mechanics, in the minds of a group of outstanding physicists, was being usurped by electro-magnetism – so much so, that it had become conceivable to him that the simplest laws of mechanics might some day be derived from the electro-magnetic theory rather than the converse.

To demonstrate that the fundamental concept of motion in mechanics might be derivable from electromagnetic considerations would be, as he remarked, a "reversal of roles" indeed. He says:

Whereas previously one had tried to explain magnetism and electricity by rotations and oscillations of the smallest parts of bodies, the present aim was to derive the fundamental laws of motion for bodies themselves from the laws of electro-magnetism.[56]

One, of course, has the impression that Boltzmann believed that such a shift in the foundations of physics, although conceivable and worth exploring, would not be feasible in the long run. Recognizing that field theory was taking the place of atomistics on one frontier, Boltzmann thought it important to recognize that

some very recently observed phenomena in cathode rays and electrolysis have led to the assumption that even electricity has an atomic constitution, consisting of discrete elements, namely electrons.[57]

The overall impression that Boltzmann conveyed while reviewing all of these cross currents in physics at the turn of the century, was that it is critically important to evaluate all clearly worked out competing mental pictures and to explore their implications for physics without too enthusiastically "overshooting the mark".

No analysis of Boltzmann's philosophy of theory construction would be complete without some reference to his profound admiration of Darwinian theory, or as he refers to it, "that most splendid mechanical theory in the field of biology".[58] As mentioned in his inaugural lecture as Mach's successor in Vienna, like Mach he held the view that

no theory is absolutely true, and equally hardly any absolutely false either, but that each must gradually be perfected as organisms must according to Darwin's theory. By being strongly attacked, a theory can gradually shed inappropriate elements while the appropriate residue remains.[59]

Essentially Boltzmann here argues in favour of an evolutionary epistemology that embraces, for scientific theories, a kind of natural selection and survival of the fittest. He says:

We understand [from the doctrine of Darwin] why it was useful and important for our species that certain sense impressions were flattering and therefore sought after, while others were repellent; we see how advantageous it was to construct the most accurate pictures possible of our surroundings and strictly to keep apart that which coincided with experience as true from what did not as false. Thus we can explain the genesis of the concept of beauty, just as that of the concept of truth, in terms of mechanics.[60]

As Boltzmann puts it, when our innate drives overshoot the mark – a characteristic element of behaviour that is biologically built into our human mechanism – then nature provides correctives and constraints that favour the survival of useful experience and the maintenance of the machinery of the world.

Harvard University

* Acknowledgment: I am happy to express my gratitude to the National Science Foundation for its support in my research on Boltzmann.

NOTES

[1] Most of Boltzmann's philosophical reflections were published in a single volume of 23 essays: Ludwig Boltzmann, *Populäre Schriften* (Leipzig, 1905). A convenient English translation of many of these essays and some other papers by Paul Foulkes, was edited by Brian McGuinness: Ludwig Boltzmann, *Theoretical Physics and Philosophical Problems. Selected Writings* (Dordrecht, 1974). These works will be referred to below as *PS* and *TP & PP* respectively. Accordingly this reference: Boltzmann, 'An inaugral lecture on natural philosophy (1903)', *TP & PP*, p. 155/6, and *PS*, p. 340/1.
[2] *Ibid.*, *TP & PP*, p. 156 and *PS*, p. 340/1.
[3] *Ibid.*, *TP & PP*, p. 157 and *PS*, p. 343. He says here: "Ich soll lehren mit sauerm Schweiss, was ich selbst nicht weiss". At the head of the Preface to Part I of his *Vorlesungen über Maxwells Theorie der Elektricität und des Lichtes* (Leipzig, 1891) Boltzmann gives another version of these lines from Faust: "So soll ich den mit sauerm Schweiss, Euch lehren, was ich selbst nicht weiss". Characteristically Boltzmannesk is what Faust actually said in Part I, Scene I (Nacht) of Goethe's tragedy; here given in its fuller version:

Drum hab' ich mich der Magie ergeben,
Ob mir, durch Geistes Kraft und Mund,
Nicht manch Geheimniss würde kund;
Dass ich nicht mehr mit sauerm Schweiss,
Zu sagen brauch was ich nicht weiss;
Dass ich erkenne was die Welt
Im Innersten zusammenhält,
Schau' alle Wirkenskraft und Samen,
Und thu' nicht mehr in Worten kramen.

See p. 30 in Goethe, *Faust. Eine Tragödie* (Stuttgart, 1838). The underlying rationale for Boltzmann's choice for these self-censuring lines is given in the Preface to the *Vorlesungen* and relates to his concern about the harsh and dry formalism that seems to be called for in order to lay bare the "true nature of electricity". He says there: "Dieses Wort Faust's ist wohl Niemandem so sehr aus der Seele gesprochen wie dem, der über die wahre Natur der Elektricität vortragen will. Möge man daher dies bischen Poesie als Einleitung zu trockenen Formeln nicht mit allzu schelen Augen ansehen".

[4] *Ibid.*, *TP & PP*, p. 155 and *PS*, p. 341: "Im Hinblick auf diese alten philosophischen Systeme möchte ich fast sagen dass man in mir den Bock zum Gärtner gemacht hat".

[5] Boltzmann, 'On the significance of theories (1890)', *TP & PP*, p. 33, and *PS*, p. 76.

[6] Boltzmann, 'The second law of thermodynamics (1886)', *TP & PP*, p. 13/4, and *PS*, p. 26.

[7] *Ibid.*, *TP & PP*, p. 14, and *PS*, p. 26.

[8] *Ibid.*, *TP & PP*, p. 32, and *PS*, p. 56.

[9] Boltzmann, 'On the significance of theories (1890)', *TP & PP*, p. 34, and *PS*, p. 77.

[10] *Ibid.*, *TP & PP*, p. 36, and *PS*, p. 79/80.

[11] Boltzmann, 'Lectures on the principles of mechanics, Part I (1897)', *TP & PP*, p. 225; *Vorlesungen über die Principe der Mechanik*, 1. Theil, Grundbegriffe, (Leipzig, 1897), p. 1/2.

[12] Boltzmann, 'On the significance of theories (1890)', *TP & PP*, p. 36, and *PS*, p. 79/80.

[13] *Ibid.*, *TP & PP*, p. 36, and *PS*, p. 80.

[14] Boltzmann, 'On statistical mechanics (1904)', *TP & PP*, p. 160, and *PS*, p. 346/7.

[15] From 'Spruch des Konfucius', Aus dem Musenalmanach für 1800, p. 304 in volume II of *Schiller's sämmtliche Schriften*, Historisch-kritische Ausgabe. Erster Theil, Gedichte, hrsg. Karl Goedeke (Stuttgart, 1871). These also were among the favourite lines of Niels Bohr, who quoted them often. See e.g., p. 11 in Léon Rosenfeld, *Niels Bohr. An Essay* (Amsterdam, 1961); and p. 226 Abraham Pais, 'Reminiscences from the post-war years', in *Niels Bohr. His Life and Works As Seen by His Friends*, ed. by S. Rozental (Amsterdam, 1967). It is noteworthy that Boltzmann dedicated his *Populäre Schriften* of 1905 to Schiller "der unübertroffener Meister der naturwahren Schilderung echter, aus tiefstem Herzen kommender Begeisterung".

[16] V. I. Lenin, *Materialism and Empirio-criticism, Critical Comments on a Reactionary Philosophy*. See p. 96/7 and 287/8 of the Abraham Fineberg translation of the Russian text of 1908, in volume 14 of Lenin's *Collected Works* (Moscow, 1962).

[17] Boltzmann, 'On the question of the objective existence of processes in inanimate nature (1897)', *TP & PP*, p. 64 and *PS*, p. 172.

[18] Boltzmann, 'On the principles of mechanics (1902)', *TP & PP*, p. 150/1, and *PS*, p. 336.

[19] Boltzmann, 'On the question of the objective existence of processes in inanimate nature (1897)', *TP & PP*, p. 68/9, and *PS*, p. 178/9.

[20] *Ibid.*, *TP & PP*, p. 69, and *PS*, p. 179/180.

[21] *Ibid.*, *TP & PP*, p. 70, and *PS*, p. 180.

[22] Boltzmann, 'On the fundamental principles and equations of mechanics (1899)', *TP & PP*, p. 105/6, and *PS*, p. 259/260.

[23] Boltzmann, 'On the indispensability of atomism in natural science (1897)', *TP & PP*, p. 41 ff, and *PS*, p. 141 ff.

[24] *Ibid.*, *TP & PP*, p. 50, and *PS*, p. 157.

[25] Boltzmann, 'On the development of the methods of theoretical physics in recent times (1899)', *TP & PP*, p. 91, and *PS*, p. 216.

[26] Boltzmann, 'On the indispensability of atomism in natural science (1897)', *TP & PP*, p. 49, and *PS*, p. 155/6.

[27] Boltzmann, *Vorlesungen über Gastheorie*, II. Teil, Einleitung (Leipzig, 1898). For English edition see Steven G. Brush *Lectures on Gas Theory* (Berkeley, 1964).
[28] Boltzmann, 'On the indispensability of atomism in natural science (1897)', *TP & PP*, p. 46, and *PS*, p. 149.
[29] *Ibid.*, *TP & PP*, p. 47, and *PS*, p. 152.
[30] Boltzmann, 'On the fundamental principles and equations of mechanics (1899)', *TP & PP*, p. 106/7, and *PS*, p. 260/1.
[31] Boltzmann, 'On the question of the objective existence of processes in inanimate nature (1897)', *TP & PP*, p. 72/3, and *PS*, p. 183–185.
[32] *Ibid.*, *TP & PP*, p. 75, and *PS*, p. 187.
[33] Boltzmann, 'On the indispensability of atomism in natural science (1897)', *TP & PP*, p. 44/5, and *PS*, p. 147.
[34] *Ibid.*, *TP & PP*, p. 46, and *PS*, p. 149.
[35] Boltzmann, 'On the development of the methods of theoretical physics in recent times (1899)', *TP & PP*, p. 82, and *PS*, p. 205.
[36] *Ibid.*, *TP & PP*, p. 83, and *PS*, p. 205/6.
[37] *Ibid.*, *TP & PP*, p. 85–87, and *PS*, p. 209–211.
[38] *Ibid.*, *TP & PP*, p. 90/1, and *PS*, p. 215/6.
[39] *Ibid.*, *TP & PP*, p. 88/9, and *PS*, p. 213/4.
[40] *Ibid.*, *TP & PP*, p. 91–93, and *PS*, p. 216–219.
[41] *Ibid.*, *TP & PP*, p. 93–96, and *PS*, p. 219–222.
[42] *Ibid.*, *TP & PP*, p. 96, and *PS*, p. 222.
[43] Boltzmann, 'On the fundamental principles and equations of mechanics (1899)', *TP & PP*, p. 104, and *PS*, p. 256/7.
[44] Boltzmann, 'On statistical mechanics (1905)', *TP & PP*, p. 172, and *PS*, p. 362.
[45] Boltzmann, 'On the development of the methods of theoretical physics in recent times (1899)', *TP & PP*, p. 97–99, and *PS*, p. 223–226.
[46] *Ibid.*, *TP & PP*, p. 97, and *PS*, p. 224.
[47] *Ibid.*, *TP & PP*, p. 100, and *PS*, p. 227.
[48] Boltzmann, 'On the fundamental principles and equations of mechanics (1899)', *TP & PP*, p. 107/8, and *PS*, p. 261/2.
[49] *Ibid.*, *TP & PP*, p. 108, and *PS*, p. 262.
[50] *Ibid.*, *TP & PP*, p. 111, and *PS*, p. 266.
[51] *Ibid.*, *TP & PP*, p. 111–114, and *PS*, p. 266–271.
[52] *Ibid.*, *TP & PP*, p. 114/5, and *PS*, p. 270–272.
[53] *Ibid.*, *TP & PP*, p. 126, and *PS*, p. 286.
[54] *Ibid.*, *TP & PP*, p. 117, and *PS*, p. 275.
[55] *Ibid.*, *TP & PP*, p. 142, and *PS*, p. 325.
[56] Boltzmann, 'On the principles of mechanics (1900)', *TP & PP*, p. 149, and *PS*, p. 334.
[57] *Ibid.*, *TP & PP*, p. 145, and *PS*, p. 329.
[58] *Ibid.*, *TP & PP*, p. 133, and *PS*, p. 314.
[59] Boltzmann, 'An inaugural lecture on natural philosophy (1903)', *TP & PP*, p. 153, and *PS*, p. 339.
[60] Boltzmann, 'On the principles of mechanics (1900)', *TP & PP*, p. 133/4 and *PS*, p. 314.

V. KARTSEV

THE MACH–BOLTZMANN CONTROVERSY AND MAXWELL'S VIEWS ON PHYSICAL REALITY

1. The first scientific paper read by young Boltzmann was Maxwell's 'Illustration of the dynamic theory of gases' (Phil. Mag., 1860). There exists evidence which suggests that since that time Boltzmann was ever influenced by Maxwell and shared his interests. Such testimony is provided not only by comparison of Maxwell's and Boltzmann's works in the kinetic theory of gases, thermodynamics, and electrodynamics, but also by a comparison of the methods of creative work applied by the scientists, and of the systems of their ideas of physical reality. Boltzmann used to say that if not for Schiller, he would be different; he might well say the same about Maxwell, and with better grounds.

2. Maxwell was greatly influenced by the philosophy of William Hamilton, and it enabled him, a profoundly religious man, to sharply delineate between the physical world which objectively exists outside an individual's mind and a 'divine' world which cannot be judged by sensual comprehension and the logic of the mind. His belief in the objective existense of physical reality perceptible to the senses also covered the sphere of physical phenomena indirectly represented by mathematical symbols and formulae. Even though this realm could not be investigated by the senses directly, its perception was so closely associated with past experiences that Maxwell felt as if it were experienced directly. He was one of those scientists who, in his own words,

> learn how fast a planet rushes in space, and this causes in them a feeling of exquisite excitement. They compute the forces with which celestial bodies are attracted, and feel their own muscles to strain from effort [1].

3. As seen from numerous facts cited in Boltzmann's biography, this scientist appreciated sensual perception to the highest possible

J. Hintikka, D. Gruender, and E. Agazzi (eds.), Pisa Conference Proceedings, Vol. II, 199–205.

extent. For example, there is his statement concerning the happiness felt by Columbus, and unattainable to other researchers, who discovered a new fact of science in the framework of sensual perception [2].

Gradually, as Boltzmann went on with his research, he replaced sensual perception with what he termed an "inner image", by which he meant some theoretically interpreted and emotionally coloured perception of a theory as a reflection of physical reality. Once Boltzmann spoke to the effect that he would be happy if someone managed to invent an inductive method comparable to the deductive one which might enable the "inner image" to appear as clearly and consistently [3].

Operation with "inner images" enables one to develop a physical hypothesis. As an example, Boltzmann said that we know nothing about the structure of the atom until we have formulated a hypothesis [4]. Here Boltzmann directly and unambiguously refers to a physical hypothesis which in the end proceeds from sensual perception.

When we emphasize the closeness between the views of Maxwell and Boltzmann, we should mention Boltzmann's deductive scheme, which is the reverse of Maxwell's: the former tried to reduce every theoretical inference to an experimentally reproducible scheme. This is how he wrote his paper on the verification of Maxwell's theoretical inference concerning the independence of the viscosity factor with respect to the density of the medium; on the verification of Maxwell's statements related to correlation between a material's refraction index and dielectric constant; verification of ideas expressed by Maxwell and Faraday on the rotation of the polarization plane of light; and others.

4. Now, were these ideas of Boltzmann merely a technique, or were they of a basic, methodological nature? Let us analyse this point using the example of Boltzmann's atomism. Boltzmann's atomism is closely associated with his sensual perception. He admitted the existence of such tiny bodies that only when they have assembled in millions do they become capable of exciting our senses. These are called atoms and molecules [5].

It is noteworthy that a proof of Boltzmann's 'H-theorem' (1872) can be obtained by a purely formal method without resort to atoms as realities. Among others, this derivation was made by N. N. Pirogov, a Russian physicist little known in the West. As Spassky puts it, Pirogov ... "is quite conscious in proceeding from an exclusively statistical interpretation of the second principle . . . ."[6]. Boltzmann's deduction relies heavily on the notions of the structure of molecules, the nature of their motion, peculiar features of their interaction, etc., whereas Pirogov's molecules may possess any shape or structure, and the laws of their interaction may vary widely. Thus, in Boltzmann's article of 1872, which is especially representative of his conception, Boltzmann's atomistic theory is ideological rather than instrumental. Even referring to physical problems which might be solved without the scheme which, according to Boltzmann, admits sensual verification, he nevertheless tries to follow this scheme.

5. Mach's biographers testify that Mach, like Maxwell and Boltzmann, was convinced of the eternal value of immediate sensual perception. Moreover, it is well known that he started research by studying the senses as an object rather than a means of investigation (in the 1870s). His experimental research in physics started later, in the 1890s. This fact, as well as keen interest in senses and perceptions not only in man but also in animals, and his interest in Buddhism, formed the basis of his constructions in philosophy which rely on perceptions as the world's prime elements [7]. Mach's philosophy, built up on the recognition of the absolute preeminence of perceptions was later termed "phenomenalism". Phenomenalists stop at a point where Maxwell and Boltzmann spread sensual perception of the real world over to the "inner image", which arises as a result of approaching physical theories and mathematical constructions as adequate means of reflecting actually existing physical reality.

6. In view of the fact that this dissimilarity in the methodological approaches to scientific activities inevitably affects their productivity, let us trace the realization of these prerequisites in the works of Maxwell and Boltzmann.

When Maxwell started with his "illustrations", he proceeded from numerous well-established experimental facts and did not doubt the discrete structure of bodies or the unordered motion of discrete particles in gases. To Maxwell, these particles were "atoms": in other words, bodies which could not be cut in half. Maxwell's "atoms" are ideally elastic balls which possess mass and velocity. When Maxwell introduced a statistical velocity distribution, he made a step unprecedented in its boldness. He replaced his former idea of atoms in deterministic relations envisioned in sensual form with that of their probabilistic relations expressed statistically. This step would have been impossible if a theory which has proved its truthfulness in some way or other were refused the properties of an "inner image", which is, in the final end, sensually processed reality.

Maxwell's conclusion concerning the independence of the coefficient of friction of a gas of its viscosity was a coronal of the probabilistic approach to the kinetic theory of gases and, in general, to the mode of describing physical reality. This inference contradicted Maxwell's earlier ideas, and experimental results; moreover, it did not agree with common sense. There arose a conflict between "sensually evident" and "theoretically reliable" knowledge. Maxwell's choice was in favour of his hypothesis concerning probable velocity distribution of molecules. We can only guess the way out of the situation which Mach would have taken as a physicist; but as a consistent philosopher, he was to rely on his sensual experience alone which, in this particular case, blocked the supply of additional objective knowledge. Later, it was proved (by Boltzmann) that experimental data disproved "sensually evident" experience and supported Maxwell's theoretical inferences. (Note that Maxwell arrived at his uncertainty about the reliability of information supplied by the senses in experimentation with "colour blindness" associated with the investigation of Daltonism.) As we see, Maxwell's methodological stand, his "style of thinking," was of extreme importance in this case.

Boltzmann accepted the kinetic theory of gases in Maxwell's statistical interpretation not only as a starting point for subsequent specialized research, but also as a milestone in the shaping of his scientific world outlook. From the idea of inevitable correlation between any theoretical construction and sensual perception, he proceeded to the firm belief in that even though perception by the senses is a mandatory component of a knowledge of physical reality, a theory verified in some way or other might also be such a component.

7. In Boltzmann's paper written in 1887 we can trace the method by which he managed to disclose the difference between the mechanical and physical representations of an equiprobable state. There is no doubt he was influenced by the provocation of "Maxwell's demon". The paradox represented in this way brought Boltzmann from the consideration of each individual molecule to the account of the properties of a statistical distribution of molecules, and to the recognition of probabilistic relations characteristic of the world of molecules being a reality. Like Maxwell, Boltzmann reached a

qualitative leap by accepting entropy as something absolute which cannot be reduced exclusively either to the mechanical or to the electrodynamical. Again, like Maxwell, he applied two major preconditions: recognition of actually existing molecules, and a hypothesis which admits probabilistic relations in the world of molecules; in other words, he used Maxwell's assumptions and methods of interpreting physical reality. M. Plank believed this development, the release of the concept of entropy from man's art of experimentation and the concurrent awarding the second principle of thermodynamics the status of a real principle, to be the main result of L. Boltzmann's activity. This step consisted in complete abandonment of a "phenomenological" entropy in favour of the probability which actually exists in the physical world. Boltzmann interpreted what was formerly unexplainable, namely Nature's attachment to irreversible processes, as a trend towards the most probable state. The introduction of a new concept of mathematical probability relies on the recognition of its denotation as a real thing, and it brought Boltzmann, like Maxwell before him, to a remarkable discovery in science [8].

Now, might Mach be prepared to accept this idea? The answer can be only in the negative. Apart from his general trend of thought, which was mentioned earlier, one cannot ignore Mach's statement to the effect that economy in the communication of knowledge and in approaching understanding is a significant feature of science. According to Mach, the major goal of science consists in finding homogeneous, consistently existing elements of Nature's processes. Mach is known to regard the so-called "world element" as such a basic component.

In Mach's opinion, science itself is to be treated as a "minimal task consisting in a possibly complete representation of facts with the least possible expenditure of mental effort" [9]. It can be seen from the above that the introduction and operation of probability as a new conception would have been utterly alien to Mach's methodology, and he could not arrive at such an important physical result while investigating his own philosophy.

8. In their apology for Mach's phenomenalism, Friedrich Adler, Joseph Petzoldt and Philipp Frank plead to the ghost of Maxwell, whom they appreciate as one of the greatest authorities in physics of all times. Frank writes unambiguously:

Were Mach's economy principle actually used as the basis in the theory of knowledge, examples of thinking provided by the leading figures of the spirit would be forgotten, the flight of their thought and, correspondingly, progress in science, would be discontinued [10].

In order to see that these lines are unsubstantiated, it is sufficient, Frank believes, to recall the views of James Clerk Maxwell on the nature of physical theories, particularly his introduction to an article on lines of force discussed by Faraday (1885). It would take little time, Frank thinks, to show that Maxwell fully shared the phenomenalistic viewpoint.

One cannot but see inadequate ground in the above discourse to place Maxwell among phenomenalists. On the contrary, while discussing physical lines of force Maxwell never failed to make it emphatic that the lines of force were not to be treated as abstractions of physics. He compared them to environmental stresses similar to those in human muscles; in other words, he admitted objectively existing reality possessing properties similar to sensual perception in their immediate perceptibility.

9. Boltzmann consciously accepted Maxwell's methodology and fought against attempts to regard Maxwell as a phenomenalist. In his analysis of the situation which had formed in physics by the rise of the XXth century, V. I. Lenin showed that the followers of Mach tried to disprove materialism by recalling that Maxwell restricts his theory to a set of equations. Boltzmann rejected this view, for he understood that it actually disproved phenomenological physics [11].

It was V. I. Lenin who demonstrated the groundlessness of the attempts by a group of philosophers (Ray et. al.) at treating both Maxwell and Hertz as "mechanists".

The greatest discoveries in the late XIXth and early XXth centuries and afterwards, such as the development of the concept of electromagnetic field, the formulation of the laws of thermodynamics, the theory of quanta, relativity theory – all these emerged within the framework of methodological ideas rooted in the concepts put forward by Maxwell and Boltzmann. In the stubborn struggle associated with the controversy between Mach's and Boltzmann's scientific and philosophical concepts, advocates of phenomenalism tried to refer to Maxwell as an authority. Yet, as analyses in history of science prove beyond doubt, it was Boltzmann who accepted the new methodological tradition started by Maxwell.

BIBLIOGRAPHY

[1] J. C. Maxwell, 'On the mathematical classification of physical quantities', in J. C. Maxwell, *The Scientific Papers*, Cambridge, vol. 2, 257–266.
[2] A. Sommerfeld, 'Ludwig Eduard Boltzmann,' *Neue Deutsche Biographie*, Bd. 2. Berlin, 1955. S. 436–443.
[3] See Ludwig Boltzmann, *Essays and Speeches*, Nauka, Moscow, 1972, p. 6.
[4] Ibid., p. 8.
[5] Ibid., p. 8.
[6] B. I. Spassky, 'On N. N. Pirogov's research in statistical substantiation of the second principle of thermodynamics', in *Collected Articles: History and Methodology of Natural Sciences*, Moscow 1970.
[7] John T. Blackmore, *Ernst Mach*, Univ. of Calif. Press, Berkeley 1972, p. 205.
[8] Ludwig Boltzmann, 'Über die Beziehung', in Boltzmann, *Werke*, Bd. II, Wien, 1909.
[9] Ernst Mach, *Mechanica*, SPB 1909, str. 409 (in Russian).
[10] P. Frank, *Between Physics and Philosophy*, Cambridge, Mass. 1941, 31.
[11] V. I. Lenin, *Complete Works*, Fifth Russian Edition, Vol. 18, p. 316.

O. A. LEZHNEVA

BOLTZMANN, MACH AND RUSSIAN PHYSICISTS OF THE LATE NINETEENTH CENTURY

Starting from the seventies and into the nineties of the last century, physicists and philosophers paid much attention to a range of problems associated with correspondence between the basic concepts of classical mechanics and the theory of thermal phenomena.

Initially, uncertainty about the eternal stability and universal nature of the fundamentals of Newtonian mechanics appeared with progress in thermodynamics; later on, some obstacles involved in relating the laws of electrodynamics to these fundamentals were understood.

Within the system "mechanics–thermodynamics", conflict started with the attempts to prove the Second Principle of Thermodynamics from the angle of traditional, classical mechanics: in mechanics, phenomena are thought to be reversible, whereas the Second Principle postulates irreversibility and unidirectional nature of actual processes. The model of the structure of matter as a sum of atoms – which was then widely accepted in mechanics – could not be substantiated with *direct* experimental results.

There were two possible ways out of the situation: first, one could ignore the existence of atoms and altogether abandon any attempt to provide specific explanations for microstructures and microinteractions, and to purposefully restrict oneself to phenomenological description based on experimental results – this is what Mach later supported with his philosophy; second, one could proceed from atomistic concepts in the recognition of causal relations between thermal phenomena and mechanical motion to search for a kinetic substantiation to the Second Principle of Thermodynamics. This was a more difficult road, but Boltzmann took it. What was the stand taken by most prominent Russian physicists?

In 1871, a master's thesis was presented at Moscow University by N. A. Umov (1846–1914) who became the first Russian theoretical physicist in the modern meaning of the term. This paper was entitled

J. Hintikka, D. Gruender, and E. Agazzi (eds.), Pisa Conference Proceedings, Vol. II, 207–209.

'Theory of Thermomechanical Phenomena in Elastic Solid Bodies' and treated electrical and mechanical phenomena, which formerly had been the realms of theories built upon dissimilar principles, in a more general form than had been done by Kelvin, Clausius and Zeiner. Proceeding from the First Principle of Thermodynamics, with a concurrent account of elastic and thermal stresses arising in the medium, Umov derived equations covering the Second Principle of Thermodynamics as a particular case alongside equations for the balance of elastic solid bodies and an equation of electrical conductivity. In his Doctoral Thesis (1874) Umov made the next step forward: he produced a general theory of the motion and distribution of every kind of energy. He pursued the goal of disclosing the relation between the distribution and transfer of energy in media and the movements of component particles, irrespective of a specific form of energy. In doing this, he introduced the concepts of energy motion, energy flux, and energy density at any point in the medium. Both in the subject matter and in the type of research, these creations of Umov are closer to Mach than Boltzmann, but Umov did not follow Mach in his philosophy.

D. D. Gulo, a Soviet historian of science who worked in the USSR Academy of Sciences Archives, came across the following statement made by Umov:

> In the sequence of deductions which are expected to lead us from the appearance available to our senses to the concealed mechanism, Newton postulated a number of links tied among one another and with the phenomena by means of clearly formulated mathematical relations. This is Newton's great discovery. It is by our will alone, and surely far from beneficial to knowledge, that thought can be stopped at one of such *intermediate* stages mistaken for the original source of the phenomenon in question.[1]

In the eighties, research in theory relating aspects of mechanics and thermodynamics was carried on in Moscow by V. A. Michelson, a disciple of A. G. Stoletov, and in Petersburg by N. N. Pirogov, the son of the great Russian surgeon. In 1883, in his paper 'The second law of thermodynamics from the viewpoint of analytical mechanics and probability theory', Michelson reached a conclusion to the effect that the Second Principle cannot be derived for irreversible processes on a "purely mechanical ground". This paper also provided an explanation in Russian of Boltzmann's concept of proving the Second Principle based on probability theory. In addition, Michelson was the first to attempt a theoretical definition of the Kirchhoff function by applying the statistical method to atomic oscillations (1887). It was in a similar way that W. Wien formulated a more exact statement of the law of energy distribution in the spectrum of an absolutely black body, which was done in 1893. It is also noteworthy that B. B. Golitsin introduced the concept of the temperature of electromagnetic irradiation (1893) which thus entered the field of phenomena to be explained in thermodynamics.

The hypothesis of the Universe's thermal death met its ardent opponent in N. N. Pirogov (1843–1891). In 1890, he set forth a unique statistical proof of the Second Principle of Thermodynamics based on a self-developed mathematical technique which contained an embryo theory of stochastic processes.[2] Several papers by Pirogov saw light in the *Journal of the Russian Physio-Chemical Society*, but, like Boltzmann's, his ideas found no support.

In the gradual positive turn towards Boltzmann's concepts which started already in the XXth century as a result of contributions made by Planck, Gibbs, Einstein and Smoloukhovsky, a positive role was played by T. Afanasyeva–Ehrenfest, whose publications appeared after 1906. Jointly with P. Ehrenfest, she disproved erroneous objections raised by Zermelo and Loschmidt, who opposed statistical explanation of the Second Principle.

Russian physicists who made a significant contribution to the progress of thermodynamics[3] and statistical physics generally disapproved of Mach's philosophy of subjective idealism. Among the active fighters against this philosophy were A. G. Stoletov and D. A. Goldgammer. Still, this philosophy aroused some interest among the less creative part of the Russian public, which was reflected in the publication of some of Mach's works in Russian between 1901 and 1909. It is common knowledge that a well-argumented criticism of Mach's philosophical concepts, as well as of those by his Russian adepts, was offered by V. I. Lenin in 1909 in his book *Materialism and Empiriocriticism.*

NOTES

[1] Gulo D. D., *Nikolai Alekseyevich Umov*. 1846–1914. Moscow, Nauka, 1971 (in Russian).

[2] Pirogov's results were carefully analysed by B. I. Spassky and V. P. Goncharov (*History and Methodology of Natural Sciences*, 1, Physics, Moscow University, 1960 (in Russian).

[3] Mention should also be made of the works by N. N. Schiller, which related to axiomatizing thermodynamics and appeared in the early XXth century.

CARLOS-ULISES MOULINES

AN EXAMPLE OF A THEORY-FRAME: EQUILIBRIUM THERMODYNAMICS

1. INTRODUCTION

The main and explicit purpose of this paper is to provide a logical reconstruction of the structure of a portion of empirical science, namely, equilibrium thermodynamics (or "thermostatics"), considering it not as a single theory but as a whole cluster of particular theories with 'family resemblances'. In order to refer to such general structures I propose to use the term "theory-frame". It is argued that the thermodynamic theory-frame has an identity of its own that should not be confused with the identity of those single theories which are subsumed under it, i.e., a theory-frame is a kind of conceptual entity lying at an intermediate level between single theories and the whole of a scientific discipline. It is plausible to think that we would find structures of a similar level of generality not only in the case of thermodynamics but in other parts of physics as well. I shall briefly argue at the end that this fact might be interesting not only for a synchronic analysis of science, but also for studying certain aspects of scientific evolution.

Besides the concrete aim of reconstructing the thermodynamic theory-frame, this paper intends to prepare the ground for a discussion of two general methodological issues: first, the relevance of previous operationalistic analyses of scientific theories to recent philosophy of science and, second, the possibility of supplementing the original version of the so-called structuralistic approach in the philosophy of science with more global analyses of the structures of science. Let me try to explain in what respect the analyses of thermodynamics as a concrete case study can impinge on the discussion of these two general issues.

Let us begin with the first general aim. The analysis of the example of thermodynamics is intended to support a general view of the relationship between qualitative operations and metrical "abstract" concepts in empirical science. And this, in turn, should provide an

J. Hintikka, D. Gruender, and E. Agazzi (eds.), Pisa Conference Proceedings, Vol. II, 211–238.

evaluation of "operationalism", both in its classical and more recent versions. It is not a casual fact that, historically, many "operationalists", from Mach through Bridgman to Giles, have shown a deep interest in the foundations of thermodynamics and have tried to ground their general philosophical outlook on their analyses of this discipline. Usual expositions of thermodynamics, even at the most theoretical level, are more explicitly concerned with the role of laboratory operations for the understanding of the theory than is usually the case in other physical disciplines. This is, I think, the main reason why thermodynamics has attracted philosophers with an operationalistic bias and has repelled or been ignored by those with an anti-operationalistic outlook.

The present analysis, however, is neither "operationalistic" nor "anti-operationalistic" – at least to the extent that these labels have a reasonably clear meaning. Operationalists rightly pointed to an important feature of empirical science – namely, the significance of qualitative laboratory operations for the adequate use of scientific concepts; nevertheless, the present analysis does by no means support the quite stronger claim that the empirical meaning of scientific concepts can be completely reduced to qualitative operations, nor even the more liberal form of recent operationalism, according to which metrical theoretical concepts are always tightly linked with qualitative operational notions by means of so-called "representation theorems".

The other general goal of this essay can be best understood if we consider the discussion of the concept of a *scientific theory.* The basic units of methodological analysis for "classical" philosophy of science were *statements.* And from this standpoint, it was natural to view theories as just sets of statements (more precisely: sets of axioms together with their consequences). Following Stegmüller's terminology in his [1976], let us call this kind of analysis "micrological". The micrological analysis of empirical science leads to the more or less explicit conclusion that science only consists of sets of statements (called "theories"). Kuhn, Toulmin, and other critics of the classical conception argued in intuitive terms that there are more general structures in science than theories in the classical sense. They looked for new terms with not very precise meaning to refer to these structures: "paradigms", "ideas of natural order", "disciplinary matrices". It first seemed to many authors that these concepts were too vague to be of any use in the logical reconstruction of science. However, Sneed and Stegmüller showed that certain kinds of structures could be formally defined that have many of the characteristics intuitively pointed out by Kuhn and others. Sneed and Stegmüller reintroduced the term "theory" for these structures. Then, to the statement that "scientific theories are something more than just sets of statements" a precise meaning could be attached. Stegmüller argued that micrological analysis had been too shortsighted and that it should be replaced by a "macrological" analysis of the more general structures of science: theories in the new sense. Concrete examples of these conceptual structures now called "theories" are: classical particle mechanics; Lagrange particle mechanics; rigid body mechanics; special relativistic particle mechanics, and simple equilibrium thermodynamics. (These are all examples that have been actually reconstructed along the lines of the structuralistic program).

Therefore, the structuralistic analysis detects an intermediate level between the whole of empirical science and individual empirical laws (or conjunctions of these). This is the level of theories in the structuralistic sense.

Now, the guiding tenet of the present essay is that the picture of science offered by the Sneed-Stegmüller macrological analysis is still incomplete in an important respect. We should proceed still "more macrologically" to complete that picture. The tenet is that there are more general structures than simply mere theories in the original structuralistic sense, i.e., general frameworks containing a number of theories and such that they can be formally distinguished from other such general frameworks. People often use such labels as "classical mechanics" "(equilibrium) thermodynamics", or "quantum mechanics", and try to say something methodologically relevant about the entities subsumed under these labels. On the other hand, these labels are not intended to cover just one particular theory in the structuralistic sense, say, classical particle mechanics as reconstructed by Sneed. On the contrary, it is presupposed that the label "classical mechanics" covers such things as classical particle mechanics, Lagrangian particle mechanics, Hamiltonian particle mechanics, rigid body mechanics, and perhaps still other theories. It is admitted that all these theories have something in common, some "family resemblances" which differentiate them from other such "families" of "classical physics". The whole group of theories is often presented in the same textbook. Students learn all of them together as a coherent whole.

Let us call any such "coherent group of theories" a "theory-frame". We would like to identify theory-frames as distinct conceptual entities, as, say, the theory-frame of classical mechanics, the theory-frame of thermodynamics, the theory-frame of quantum mechanics, etc. Kuhn's notion of "normal science" probably applies better to the kind of activity developed within such general theory-frames than to just the work done within a specific theory. And, perhaps, we could also speak of something like "paradigmatic exemplars" for theory-frames: single theories that are "paradigmatic" for the construction of the rest of theories within a given frame. For example, we could say that, in a certain sense, classical particle mechanics is a "paradigm" for classical mechanics as a whole and simple equilibrium thermodynamics a "paradigm" for the thermodynamical frame.

Now, of course, the question arises whether something reasonably precise and methodologically interesting can be said for these intuitively grasped theory-frames. This essay attempts to support an affirmative answer to the question. The starting point will be that we intuitively accept the existence of such thing as "the theory-frame of (equilibrium) thermodynamics" and the final result will be to provide a precise definition of this notion and to demonstrate its methodological significance.

II. MATHEMATICAL NOTIONS

Besides the usual notions of set theory and arithmetic, more special concepts and notations will be used.

1. If N is a set, then $k(N)$ denotes its cardinality.

2. So-called "set-theoretic *projection functions*" π_i can be defined on ordered n-tuples picking out the i-th term of any tuple. That is, the domain of a projection function π_i is the set of all ordered tuples and its range is the set of all entities which occupy the i-th place in any such tuple: If $x = \langle y^1, \ldots, y^i, \ldots, y^n \rangle$, then $\pi_i(x) = y^i$.

In order to simplify the notation and to make the formulae more perspicuous, I shall not *use* projection functions explicitly in this paper, but I shall *assume* them implicitly. Let us admit the following convention. If $x = \langle a^1, \ldots, a^i, \ldots, a^n \rangle$, then we shall write a^i_x instead of $\pi_i(x)$.

3. Let P be any (n-ary) predicate; then $\hat{P}$ is its extension, i.e. the set of all entities satisfying P.

4. Let R be any binary relation: $D_{\text{I}}(R)$ is its domain and $D_{\text{II}}(R)$ its range. We assume the notions of equivalence and ordering relations as well-known. Further, we need the following less well-known kind of relation: R is *conditionally connected* iff, for all x, y, z, if xRz and yRz, then xRy or yRx.[1]

5. For a clear and precise formulation of some statements, logical symbols might be useful. For the connectives: $\neg$ (not), $\vee$ (or), $\wedge$ (and), $\rightarrow$ (if . . . , then . . .), $\leftrightarrow$ (iff). For the universal quantifier: $\bigwedge x \ldots$, and for the existential quantifier: $\bigvee x \ldots$. However, use of these symbols will be made only in those cases where it really seems advisable. For simple statements, the corresponding English words will be used instead. When there is no danger of confusion, we shall also make use of the convention of writing free variables for universally quantified variables.

6. If f is a monadic function, we write Df for its first derivative. If f is an n-adic function, and x_i one of its arguments, we write $D_{x_i}f$ for the partial derivative of f with respect to x_i.

To express that the function f is held constant in a neighborhood of its argument x, we write $\Delta f(x) = 0$.

If f has a relative maximum in x, we write $rmax(f/x)$; and if it has a relative minimum, $rmin(f/x)$.

III. GENERAL METHODOLOGICAL REMARKS

1. Let θ be any theory, and M_θ be the set of models of θ. A way of characterizing this set is to formalize θ and choose adequate interpretation functions onto certain previously determined domains. However, as is well-known, there is a more expeditious way of characterizing M_θ: This is the (informal) method of defining a set-theoretic predicate to axiomatize θ. Let us call such a set-theoretic predicate P_θ. The axioms of θ are just the definitory conditions for P_θ. Then, we simply have: $M_\theta = \hat{P}_\theta$, i.e. a model of θ is any entity satisfying P_θ.

The method of defining a set-theoretic predicate has been widely used by Suppes and others to axiomatize many empirical theories.

There is no reason to doubt that this method can also be applied to the rest of empirical theories to which it has not been applied so far, at least to theories in mathematized science. Consequently, I shall make the idealized, but admissible assumption that it has already been applied to the axiomatization of at least the thermodynamic theories which are of our concern here.

2. Every model x of such an axiomatized θ will be a structure, that is, an ordered tuple of the form $x = \langle a^1, \dots a^n \rangle$, where the terms a^i will be simple sets, relations or functions (the primitive concepts of the theory). Their structural properties and interconnections with each other are stated in the definitory conditions (axioms) for P_θ. Now, two different models of θ, say x and y, will be structures including different terms. However, we can put the terms of both structures in a one-one correspondence to each other, according to the place they take in the tuple. For any place i, a^i_x and a^i_y will have the same formal axiomatic properties, although different empirical interpretations. Following Stegmüller [1976], we can say that a^i_x and a^i_y belong to the same "abstract a^i-concept". When physicists speak about the properties of, say, *the* mass function or *the* entropy function in general, they normally refer not to a particular mass function or a particular entropy function in a particular model of mechanics, *viz.* thermodynamics, but to the "abstract mass concept", *viz.* the "abstract entropy concept", to which the concrete masses, *viz.* entropies belong.[2] Such abstract concepts, in general, cannot be determined extensionally, but only intensionally. Some philosophers may feel uneasy about this. But I think one may use intensional concepts when there is no clear way to find adequate corresponding extensional concepts – as is the case here. Once we have accepted the former notion of an abstract a^i-concept, we might introduce the following convention: whenever a term appears without any index denoting the particular model to which it belongs (*i.e.* a^i instead of a^i_x, for example), it shall be understood as denoting the corresponding abstract concept. This convention allows us to simplify many formulations without losing formal rigor.

3. The method of defining set-theoretic predicates has been used mainly to systematize and clarify the structure of single *theories*. However, I see no reason to restrict its application to this case. It can also be used to clarify *metatheoretical* structures. To make this point clearer, let us introduce some conceptual distinctions.

Theoretical activity happens at different levels. First, there is the direct theoretization of natural phenomena that leads to the formulation of empirical concepts and laws allowing for explanations and predictions of phenomena. Let us call this stage of theoretical activity 'Conceptual Level 1'. (The stage of gathering experimental and observational data could be called 'Conceptual Level 0'). The concepts and laws of Conceptual Level 1 are loosely connected with each other and are usually expressed in rather vague terms (even if they may be mathematized). The professional scientist normally works at this level.

Some people – mainly logicians, mathematicians, and philosophically-minded scientists – sometimes feel dissatisfied with the way science appears at Conceptual Level 1 and try to bring more clarity and order into it. The objects of study are, then, not natural phenomena, but rather the concepts and laws referring to those phenomena. Let us say that this kind of study happens at Conceptual Level 2. A typical task for this level is to axiomatize an empirical theory; this can be done by introducing set-theoretical predicates.

However, our theoretization drive need not stop at Conceptual Level 2. We may try to develop general concepts and theories to account for the empirical concepts and theories that have already been clarified at Level 2, especially by means of the axiomatic method. Our objects of study are then whole clusters of axiomatized theories. This is a typically philosophical task. Let us call this stage Conceptual Level 3. The methodology of Conceptual Level 3 can be as rigorous and exact as that of Conceptual Level 2. In particular, we can also use the axiomatization by means of a set-theoretic predicate to clarify matters at Conceptual Level 3. The entities satisfying such set-theoretic predicates at Level 3 will not be empirical models; they will be rather, whole theories or entities of a similar kind. But the standards of procedure will be much the same. In other words, the way we axiomatize philosophical metatheories will be analogous to the way we axiomatize empirical theories.

The present study belongs to Conceptual Level 3. Its aim is to develop a metatheory of thermodynamic theories; and the method is set-theoretic axiomatization.

4. The metatheory propounded in this paper does not deal with all aspects of a thermodynamic theory one should consider to get a complete picture. Only those aspects and structures are considered which appear as explicable in a reasonably rigorous way at the

present stage of general methodology. In particular, this paper does not deal with all aspects of so-called "physical semantics" for thermodynamic theories.[3] The main reason for this is that there seem to be issues of physical semantics which cannot be formally reconstructed as yet. Consequently, I do not pretend to offer a complete metatheory of thermodynamics, but I certainly intend to reconstruct at least some essential aspects common to all thermodynamic theories. This means that the metatheoretical predicates introduced in this paper are, in some sense, incompletely determined, *i.e.*, determined only up to certain (essential) necessary conditions. For example, the predicate "is an equilibrium thermodynamic theory" should actually be read as "is a *possible* equilibrium thermodynamic theory". The same goes for the other main metatheoretical predicates introduced here. Or, alternatively, we could say that those predicates are completely determined, but only so far as they only refer to the formally describable aspects of thermodynamic theories. This proviso should be borne in mind when considering the formal definitions in this paper.

IV. THE CONCEPT OF AN EQUILIBRIUM THERMODYNAMIC THEORY

1. The entities which fit the equilibrium thermodynamic theory-frame (abbreviated as "E. T. T. F.") will be theories of a special sort. Let us call them equilibrium thermodynamic theories. So, our first task is to define the set-theoretic predicate "is an equilibrium thermodynamic theory" (abbreviated as "*ETT*"), which they should satisfy. Their satisfying this predicate is a necessary, but not sufficient condition for them to belong to the E. T. T. F. Something more is needed to complete the frame: the relation of such theories to certain operational structures. To this we shall come in the next sections. Let us first determine the predicate "*ETT*".

2. An example of an equilibrium thermodynamic theory has already been formally reconstructed along the lines of Sneed's metatheory: simple equilibrium thermodynamics.[4] This is the simplest full thermodynamic theory we can think of. Besides energy and entropy, the only primitive parameters considered in this theory are volume, pressure, and mole numbers. Other equilibrium thermodynamic theories are more complex. They include other parameters such as

elastic strain, electric charge, or magnetic moment. However, it is plausible to think that simple equilibrium thermodynamics should be taken as a paradigm for the other, more complex theories in the sense that the essential structure will be the same. If this is correct, then the essential features of simple equilibrium thermodynamics should be abstracted as a base for defining the general predicate "*ETT*".

The features of simple equilibrium thermodynamics which resulted from the logical reconstruction above-mentioned and which are likely to be relevant to other theories of the same frame, are the following:[5]

(a) The domain of any model of the theory is a set of states isomorphic to a continuous sequence of real numbers, such that differentiable functions can be defined on this set.

(b) A particular subset of the set mentioned in (a) is the set of all so-called equilibrium states.

(c) Every model of the theory is a structure consisting of some metric functions, among them energy and entropy, in addition to the domain and the subset of equilibrium states.

(d) The only thermodynamic-theoretical concepts are energy, entropy, and the monadic property "is an equilibrium state". All other functions are provided by non-thermodynamic (presupposed) theories (e.g. those included in physical geometry or classical mechanics).

(e) The fundamental law of the theory is a general functional relationship between entropy, energy, and other metric functions (in the case of simple equilibrium thermodynamics, these additional functions are: volume and mole numbers). This law is valid for equilibrium states z only, and it is expressed by means of an equation of the sort:

$$S(z) = f^S(U(z), \ldots)$$

or, alternatively,

$$U(z) = f^U(S(z), \ldots),$$

where f^S, viz. f^U, are general unspecified functions fulfilling certain formal properties and determining the relationship between S, *viz.* U, and the other functions appearing within the domain of f^S, *viz.* f^U.[6] Let us call all the functions so interrelated in the fundamental law "*fundamental parameters*". We shall see later – when we come to discuss the theory-frame – that all these fundamental parameters are extensive, i.e., they have the property of additivity. But this need not be of our concern for the time being.

(f) Different sorts of models (i.e., applications) of the theory result from different specifications of f^S, expressing as many different "special laws" of the theory.

(g) For a complete reconstruction of the theory, so-called "constraints" as inter-connections between different models have to be established.[7] There are two sorts of constraints: general and special constraints. General constraints are valid for any pair of models; special constraints are valid only in certain specializations of the theory. Here, we are interested only in general constraints. In Moulines, [1975], eight general constraints were postulated for the notions of equilibrium, entropy, and energy. (Other functions are supposed to be constrained in underlying theories). Gibb's "minimax" principle for entropy and energy was included among them. Among general constraints, a further distinction can be made which is essential for the present reconstruction: the statement of some constraints only involves concepts appearing in the structures defining the theory's models; but for other constraints, some qualitative operations must be mentioned that are not components of the models themselves – a well-known example is the notion of combination. The distinction was not made in previous applications of the structuralistic approach, since in this approach there is no place for operational notions. Here, however, we can, and should, make the distinction. We shall say that the first kind of constraints belong to the notion of a thermodynamic theory (something that does not involve any operational concepts). The second kind of constraints, on the contrary, are part of the conditions determining the more general structure we call "thermodynamic theory-frame". The only constraints of the first kind relevant to thermodynamics are so-called "identity-constraints". The rather more interesting and complicated constraints involving such notions as combination, partial equilibrium, and transition, belong to the second kind.

3. One of the main metatheoretical hypotheses of the present essay is that features (a) to (g) have the same form for all thermodynamic theories. This assumption implies, among other things, that the concepts of equilibrium, entropy, and energy are common to all of them (this is why their constraints are assumed to be the same throughout the whole range of theories), and that any such theory

includes, besides entropy and energy, *at least one* more fundamental parameter. This additional parameter will always be thermodynamic-non-theoretical (by feature (d)). For example, the thermodynamic study of electrostatic *viz.* magnetic systems requires the electric charge *viz.* the magnetic flux to be added as fundamental parameters;[8] for solid systems we may introduce the elastic strain;[9] moving systems require momentum and angular momentum as additional parameters.[10] All these parameters are provided by theories not presupposing thermodynamics: electrostatics, magnetostatics, rigid body mechanics.

4. The models of a thermodynamic theory are determined by the specific S and U functions and by the other fundamental parameters that are used to describe the thermodynamic systems (= sets of states) dealt with in the theory. In other words, these parameters (including S and U) are primitive concepts in the set-theoretic axiomatization of the theory. However, they are not necessarily, the *only* primitive (non-defined) functions. (Fundamental here is not used in the sense of "non-defined"). Besides the fundamental parameters, other functions are used in thermodynamic theory. Examples of such additional functions are: temperature, pressure, chemical potential and electrostatic potential. These functions are related to pairs of fundamental parameters by way of partial derivatives. Let us call them, for this reason, "*derivative functions*". For example, the temperature is said to be the partial derivative of the energy with respect to the entropy; the (negative) pressure is the partial derivative of the energy with respect to the volume, and so on. Some of these correlations (as in the case of temperature and chemical potentials) may be considered as *definitions*, so that the derivative concepts introduced (temperature, chemical potential) will also be "derived" in the sense of not being primitive. But this need not be the case for all derivative parameters of the theory. In other cases (for example, in the case of pressure, as it has been argued in Moulines, [1975]), the correlation between the derivative parameter and the corresponding pair of fundamental parameters should be taken as a statement of fact, *i.e.*, as an empirical law correlating three primitive concepts. Consequently, *at least in some* thermodynamic theories, some derivative functions will be primitive concepts, so that the models of the theory will include non-fundamental parameters besides the fundamental ones.

For the following, a derivative parameter will be understood specifically as a metric function of state which is related (either definitionally or empirically) to a couple of fundamental functions through partial derivatives.

Another primitive concept which may be introduced to describe the models of a thermodynamic theory is a numerical index set. Some thermodynamic functions may be split into different "components". For example, the mole number function in simple equilibrium thermodynamics may be split into different "partial" mole number functions for each of the chemical substances composing the total system. The same should be true for the components of the elastic strain tensor in the case of thermodynamic systems with elastic deformations.[11] In all such cases, it is more adequate not to split the primitive function (mole number, elastic strain) formally, but to introduce instead a set of index numbers – each number referring to a particular component.[12]

6. Keeping the foregoing remarks in mind, let us define the concept of an equilibrium thermodynamic theory by means of a set-theoretic predicate. In the following, θ is a variable for a "theory".

D1: θ is an *ETT* iff: For every $x \in M_\theta$ there are three sets Z, E, I and functions S, U, and a set of functions $\{f^1, \ldots f^n\}$ with $n \geq 1$, and a (possibly vacuous) set $\{g^1, \ldots g^m\}$ such that:

(1) $x = \langle Z, E, I, S, U, f^1, \ldots f^n, g^1, \ldots g^m \rangle$

(2) Z is a set of states isomorphic to an interval of real numbers.

(3) E is a subset of Z and $E \neq \emptyset$.

(4) I is a finite, possibly vacuous set of natural numbers.

(5) S, U, and every f^i is a metric function from Z into $\mathbb{R}^m$, where $m \geq 1$. At least S and U are twice differentiable.

(6) $\{g^1, .. g^m\}$ is a (possibly vacuous) set of derivative functions with respect to S, U, and $f^1, \ldots f^n$.

(7) There is a real function f^S from a subset of $\mathbb{R}^{n+1}$ into $\mathbb{R}$ such that:

(a) f^S is at least twice differentiable.

(b) f^S is a strictly monotonic function of its arguments.

(c) S is weak monotonic with respect to f^S.

(*d*) For any $z \in Z$:

$$z \in E \text{ iff } S(z) = f^S(U(z), f^1(z), \ldots f^n(z)).$$

(8) The number n is the same for all $x \in M_\theta$.
(9) Only E, S, and U are θ-theoretic predicates.
(10) For any pair $x, x' \in M_\theta$ and any state z:
  (a) if $z \in E_x$ and $z \in Z_{x'}$, then $z \in E_{x'}$.
  (b) if $z \in Z_x \cap Z_{x'}$, then $U_x(z) = U_{x'}(z)$.
  (c) if $z \in Z_x \cap Z_{x'}$, then $S_x(z) = S_{x'}(z)$.
(11) Different sorts of applications (models) of θ result from different specifications of f^S (different special θ-laws).

7. The following point about entropy and energy should be noted here. We have said that entropy and energy are common to all thermodynamic theories, but the previous definition of *ETT* does not include a stringent characterization of entropy and energy; the conditions stated for them are too general and weak in order to distinguish them in a precise way from other functions. Therefore, it could be objected that the assertion that entropy and energy are common to all thermodynamic theories does not have much content. This is in a certain sense true up to this point. D1 does not tell us much in the way of what we could call the physical semantics of entropy and energy – only some general formal properties and the statement that they are θ-theoretic in every $\theta \in \widehat{ETT}$. But there are two points to be remarked on this issue. First, I have already said that I do not pretend to give a complete semantics for thermodynamics in this paper – and this is particularly true for the concepts of entropy and energy; but even so, the assertion that entropy and energy are common to all thermodynamic theories is not devoid of meaning; this assertion is crucial in order to characterize the thermodynamic theory-frame formally—even if we have to wait for future research to give a complete semantic characterization of entropy and energy. Secondly, in the last section of this paper the attempt is made to give a closer characterization of a *part* of the semantics of entropy and energy, namely by connecting these metric functions with some special empirical operations. So, the characterization of entropy and energy given in this paper will be, after all, not so loose as one might suspect at first sight.

8. The states of systems a particular thermodynamic theory θ deals with are those states which appear in at least one model of θ. It is convenient to have a special symbol for the set of all these states:

$$\mathcal{T}_\theta = \bigcup_{x \in M_\theta} \pi_1 x$$

V. OPERATIONAL BASES FOR THERMODYNAMICS

1. Operational bases are essential for the task of characterizing the thermodynamic theory-frame. A given theory will not be considered as belonging to the thermodynamic theory-frame unless it is linked with a set of operational bases corresponding to each of the primitive functions of the theory. The operational bases determine the operational meaning of a theory, that is, they determine the way the theory is going to be applied operationally. After introducing the corresponding operational bases we know, for each function, what kind of laboratory operations are associated with it.

2. Since the foregoing paragraph, and in general the introduction of operational bases, may sound as operationalism, a few remarks are in order here. First, the introduction of the present notion of an operational basis does not mean any commitment at all to the acceptance of operationalism as a philosophical doctrine. It is true that some authors who have studied the properties of what I call operational bases (Falk, Jung, Giles, Duistermaat, Hornix) adopted operationalism or at least showed a certain operationalistic bias. But this is of no concern to me. One can use some ideas and results of certain authors without being committed to the whole of their philosophical view.

Secondly, to say that operational bases provide the operational meaning of the concepts of a theory does not at all imply either that *all* meaning is operational or that operational meaning is the only kind that matters. The semantics of physical concepts is a very complicated issue. And I am quite certain that it involves much more than laboratory operations. But, in any case, it *also* involves laboratory operations. The operational entities are an important component. In the next paragraphs I shall try to show why this is so.

Thirdly, to say that the primitive metrical functions of the theory are associated with operational bases does not mean that they are *defined* by them, nor does it mean that operational bases are epistemologically prior to the corresponding metrical concepts. It is not assumed that operational bases are "purely observational"; they are also "theory-laden" in the sense that their introduction goes hand in hand with the introduction of the primitive metrical concepts of the theory. And their meaning varies from theory to theory.

After these remarks, it should be clear that the present approach is not operationalistic – at least not in the current sense of "operationalism" – , even if I borrow some ideas from such an operationalist as Giles. All that is being asserted is that certain laboratory operations have something to do with the empirical use of certain concepts of the theory; and since we know, at least since Wittgenstein, that the use of a concept has something to do with its meaning, we can conclude that operations have something to do with the meaning of the corresponding concepts. If this is still to be considered operationalism, it is a rather watered-down sort of operationalism. And, after all, such general philosophical labels are rather unimportant.

3. What can be positively said about operational bases is that the concepts they contain are non-metrical and qualitative; they are related to qualitatively describable situations found or arranged in the laboratory. They refer to such things as different sorts of qualitatively identifiable states of matter and passages from one state to another.

4. An operational basis is a structure composed of a set of states and certain relations defined on this set. In the present case of equilibrium thermodynamics these relations are: two dyadic relations called, respectively, "transition (under isolation)" and "equilibrium", and an operation (*i.e.*, a tryadic relation) called "combination". Different operational bases will, in general, contain different kinds of transition, equilibrium, and combination relations; that is, the set of axioms implicitly defining those relations admits many different empirical models. Intuitively, this means that different operational bases express different kinds of possible transitions, combinations, and equilibrium situations. Some of them will be interesting for thermodynamics. In sum, the four primitive concepts of thermodynamic operational bases we choose here are: *state*, *transition*, *equilibrium*, and *combination*. These are the basic operational concepts of the thermodynamic theory-frame.

At this point, the question of the justification of my choice might be raised. The justification is, in general terms, the following: all that can be stated in thermodynamics about the *operational* aspects of its theories can be reduced to statements about certain sorts of combinations of states, transition processes, and states of equilibrium. This is a substantial claim about the operational aspect of the thermodynamic theory-frame, and it is difficult to see how it could be proved or refuted formally. It is a general metahypothesis about the way thermodynamic theories relate to empirical operations. Its support can only consist in actual expositions of thermodynamics and laboratory accounts. I think there exists such support. In any case, this is the hypothesis assumed here.

5. Most foundational studies in thermodynamics (Falk & Jung, Hornix, Giles) have also taken the above-mentioned concepts as basic—although for possibly the wrong reasons (operationalism). Giles and Duistermaat explicitly introduce "state", "transition", and "combination" as the primitive concepts of their reconstructions. Notice, however, that their interpretation of these concepts, especially of the concept of transition, is much more restricted than the one adopted here. (Actually, Giles only considers adiabatic transition; and Duistermaat is only interested in adiabatic and energetic transitions). Giles tries to get the concept of thermodynamic equilibrium as a defined concept, but his interpretation of it is too narrow for our present needs, since it only refers to complete (total) equilibrium. In taking a general concept of (unspecified) equilibrium as a primitive concept I follow Tisza's intuitive remarks about the concept[13] and also the expositions of Falk and Jung, and Hornix.

The present interpretation of those operational concepts is more akin to the one offered by Falk & Jung than to Giles' approach, although it does not coincide with either. In general, the present characterization deliberately attempts to be much broader and more "underdetermined" than the previous ones. However, the important point to notice here is that all these different approaches, in spite of the differences of detail, broadly coincide in the choice of the fundamental operations of thermodynamics; I believe this is because they all point to the same aspect of the thermodynamic theory-frame: its "operational bases".

6. Let us explain the intuitive meaning of each one of the four concepts integrating an operational basis. The most elusive concept is certainly the concept of a *state*; states are the fundamental entities thermodynamic theories talk about. Although in particular cases every experimental thermodynamicist knows how to distinguish one thermodynamical state from another one, it is difficult to give a generally valid characterization of the concept. Giles proposes the following "operational" criterion for identity of states:

The state of a system represents its method of preparation: two systems are in the same state if they have been prepared in the same way, or more precisely if our information about the method of preparation is the same in each case" (Giles, [1964], p. 17).

This explication is apparently simple, but it is not really workable, since it relies on two unexplicated notions: "system" and "method of preparation". Besides, whatever the phrase "methods of preparation" may mean, it seems too restrictive to identify states with methods of preparations (human manipulations of an apparatus?). It is preferable to take states as the most primitive kind of entities in thermodynamics, so that, by their very notion, they cannot be explicated in thermodynamically more primitive terms. This is also the point of view adopted by Falk & Jung:

Bei einer axiomatischen Beschreibung spielt der Zustand die Rolle eines (nicht näher zu erklärenden) Grundobjektes der Theorie: Es wird nur vorausgestzt, dass von Zuständen z als unterscheidbaren Objekten gesprochen werden kann" (Falk & Jung, [1959], p. 121).

Let us only add a negative remark to this. A state should not be visualized as *one* physical (material) object, but rather as a *class* of (thermodynamically equivalent) objects. States are not spatiotemporally localized. In this way, we can express, for example, that a particular portion of a gas in a container in laboratory L at time t is *in the same state* z as another particular portion of that gas in another container in laboratory L' at time t'. We could try to define states as certain equivalence classes of material objects; but I am not going to pursue this line here.

The notion of a state is certainly closely related to the notion of a *system*. We speak of "states of a system". One might think that it would be convenient also to introduce the notion of a system into the set of basic concepts constituting an operational basis. However, this seems to be superfluous. Following Giles, we can define a system as just a set of states. This is all we need. It is true that many textbooks of thermodynamics appear to interpret "system" as something more than a "mere" set of states; say, not only the states of a gas in a container, but the container itself. However, such material and spatial representations are misleading and out of place in a systematic exposition of thermodynamics.

7. *Transitions* are relations between states. The symbolism for it will be: $z \Rightarrow z'$. There are many different kinds of possible transitions. In general, the concept of a transition is linked to the concept of a

physical *process*. Whenever we analyze a physical system in terms of states, we can say that a given process in this system is a sequence of transitions from one state to another one. Thermodynamics is not interested in all kinds of possible transitions. Thermodynamically interesting transitions are only those that can be performed under a given sort of *isolation*. This is why some authors, notably Hornix, have also called "isolation" the kind of relation I here call "transition". Indeed, the most appropriate label to express the intuitive idea would be: "transition-under-isolation-of-the-kind-k". But this is too long a term; so we speak only of "(thermodynamic) transition". Hornix listed the four main types of thermodynamically interesting transitions (or "isolations"): transition under adiabatic isolation; transition under energetic isolation; transitions under a dimensional isolation (*i.e.*, isolating the system in a given spatial dimension); transition under a chemical isolation (*i.e.*, isolating the mass of a given chemical compound).

Each of these transitions is conceptually linked to a thermodynamic metrical function in a certain specific way. This link will be explicated in the next section. Here, the task is to analyze the general logical structure of such transition relations independently of the metric functions.

Intuitively, the concept of a transition is associated with the idea of a *temporal* sequence. If a "process leads z to z'", we naturally think that z' "takes place" *after* z—that is, we think z and z' are separated by a lapse of time. However, we should not suppose that this intuitive visualization is actually required for the formalism of (equilibrium) thermodynamic theories; in particular, we should *not* presuppose that, when talking about thermodynamic transitions, we are assuming a metrical time parameter t among the basic concepts of the theory. (Metrical) time is not essential to equilibrium thermodynamics. This is sometimes expressed by saying that we should not speak of "thermodynamics" but rather of "thermostatics".

The present concept of a transition is very broad; it includes not only "spontaneous" or "natural" transitions (as in Giles and Duistermaat), but also any kind of "manipulated" or "forced" sequential changes taking place in a system. Normally, transitions occur as a consequence of some sort of *interaction* of the system with its "surroundings". The intuitive notion of an interaction is the conceptual link between transitions and another basic operational concept:

combinations of states. Interesting combinations normally (but not always) produce transitions.

8. A *combination* is an operation in the algebraic sense on the set of states. We write: $z = z' \otimes z''$ for "state z is the result of combining states z' and z''". It could also be called "concatenation" or "composition" or "join" or "coupling", etc. (Giles and Duistermaat use the not very appropriate terms "sum" and "addition"). Combinations appear not only in thermodynamics, but in practically all branches of empirical science. Thermodynamics is interested in certain kinds of combinations: again, those which are related to thermodynamical metrical parameters. (For example: the combination of two states, whose corresponding material realizations are separated by a movable wall; or mixtures of gases).

There is no general agreement among different authors on the intuitive interpretation of thermodynamic combination. At one extreme, we find Giles taking it as a purely conceptual issue:

> It should be emphasized that the process of forming the union [combination] $A + B$ is a purely conceptual one; in particular, it does not imply any interaction between A and B (Giles, [1964], p. 22, Note).

At the other extreme, Duistermaat remarks:

> The act of joining states of physical systems and the observation of the behaviour of the coupled system is so common in experimental physics that a structure of addition could be expected in almost every physical theory. Although in many cases there is no danger of confusion, sometimes more care is needed in the physical interpretation of the concept of addition. In particular, addition of states is not considered as a mere 'mental act' but as a real physical operation.
>
> For instance, the description of the state 'two charged particles together in space' is not to be considered as a mere addition of the descriptions of both (isolated) particles. The description of this state is incomplete if it does not mention the auxiliary equipment and the effort needed in order to bring the charged particles together (Duistermaat, [1968], p. 33–331).[14]

An intermediate position seems to be the one adopted by Falk and Jung:

> *Beim Begriff der Zusammensetzung... trifft in vielen Fällen die Vorstellung des Nebeinander-Legens von Systemen und des Einschaltens einer Wechselwirkung über eine geeignete "Wand" recht gut das Gemeinte* (Falk & Jung, [1959], p. 159–160).

This characterization appears to be in better agreement with Duistermaat than with Giles. Notice, however, the compromising qualification "in vielen Fällen". It seems to mean that, although physical interaction is the normal situation when we speak of combination, the 'purely conceptual' joining of states could also be included under the general combination concept.

In any event, this will be the intuitive interpretation of the $\otimes$-operation propounded here: Any kind of joining will be subsumed under the label "combination of states". It does not matter whether this joining is interpreted as a "purely conceptual" act or as one involving physical manipulations—as long as it is thermodynamically significant, that is, as long as this particular interpretation of $\otimes$ is relevant to the determination of a particular thermodynamic function in which we are interested. When we come to discuss the formal properties of the thermodynamic theory-frame we shall see more clearly what is meant by an $\otimes$-operation being relevant to a thermodynamic function.

The only general restriction introduced here on the $\otimes$-concept is that it shall be an algebraically *closed* operation. That is, if two states whatsoever z, z' are combined within the context of a theory θ, then $z \otimes z'$ (whatever the particular interpretation of $\otimes$) is also a state in this theory. Formally, if $z, z' \in \mathfrak{T}_\theta$, then also $z \otimes z' \in \mathfrak{T}_\theta$. Possible combinations of states which would not satisfy this (an extreme example: instantaneous explosions due to combination) are excluded from the range of possible interpretations of $\otimes$.

9. *Thermodynamic equilibrium* is a dyadic relation between states. We write "$z \sim z'$" for "z is in equilibrium with z'". This relation is not reflexive, since some states are non-equilibrium states (i.e., they are not in equilibrium with themselves). But it is not irreflexive either, since some states actually are in equilibrium with themselves: they are equilibrium states.

Again, as in the case of transitions and combinations, there are many different kinds of thermodynamically significant equilibrium relations: thermal, chemical and pressure equilibrium are among the most usual forms. This means we are considering *partial* equilibrium: equilibrium with respect to a specific thermodynamic property. For example, two states can be in a chemical, but not in a thermal equilibrium. The specific kind of equilibrium attained between two different states depends on the specific kind of "walls" separating them— that is, on the specific kind of laboratory interaction performed in their combination. So, there is also some operational relationship between the concepts of equilibrium and combination.

The concept of complete equilibrium, as opposed to merely partial equilibrium, is a concept introduced by each thermodynamic theory in order to formulate its fundamental laws appropriately. As we have already seen in Section IV above, (complete) equilibrium is a typical θ-theoretical concept for each theory θ of equilibrium thermodynamics; it is not an operational concept. However, there is certainly a link connecting (not *defining*!) complete equilibrium with partial equilibrium. Roughly, the link is this: for a given theory θ, we must assume that a given state z in θ is a (complete) equilibrium state if it is in partial equilibrium with itself for all (partial) equilibrium relations introduced by the operational bases underlying θ. The formulation of this connection will be seen more clearly in the next section.

The (partial) equilibrium concept is the operational counterpart of what has been called a "derivative function" in thermodynamics (temperature, pressure, etc.). It also plays an essential role in the formulation of "quasi-dynamical" processes—in connection with Gibbs' principle or related assumptions.

10. After the foregoing informal remarks, we are now in a position to introduce the general concept of an operational basis for thermodynamics formally.

D2: B is an *operational basis* for an equilibrium thermodynamic theory iff: There are $\mathcal{T}, \Rightarrow, \otimes, \sim$ such that:

(0) $B = \langle \mathcal{T}, \Rightarrow, \otimes, \sim \rangle$.

(1) $\mathcal{T}$ is a non-empty set.

(2) $\Rightarrow$ is a dyadic relation on $\mathcal{T}$, and $\Rightarrow$ is transitive.

(3) $\otimes$ is a dyadic function from $\mathcal{T} \times \mathcal{T}$ into $\mathcal{T}$, and $\otimes$ is non-idempotent, commutative, and associative.

(4) $\sim$ is a dyadic relation on $\mathcal{T}$, and $\sim$ is symmetrical and transitive.

(5) For any $z, z', z'' \in \mathcal{T}$:
If $z \otimes z'' \Rightarrow z' \otimes z''$, then $z \Rightarrow z'$.

(6) For any $z, z', z'' \in \mathcal{T}$:
$z \otimes z' \sim z''$ iff ($z \sim z''$ and $z' \sim z''$).

Conditions D2-(1) to D2-(4) are the "structural axioms" for the notion of an operational basis; they determine the formal properties of its four components. The elements of the set $\mathscr{T}$ are called *states*. Intuitively, $\mathscr{T}$ is any set of operationally distinguishable entities of the sort we would call "states of a system". Nothing is said about the cardinality of $\mathscr{T}$. This question, which is important when we come to deal with models of thermodynamic theories, is left open here—for the very reason that we are dealing with the *operational* aspects of thermodynamics; the question of cardinality is a *theoretical* (non-operational) issue.

The only formal property we require of transition is that it be transitive; that is, if the system can go from z into z' in a certain way, and z' can go into z'' in the same way, then it should be possible to say that z goes into z'' in *this* way—without mentioning z'. Notice that it is neither required nor ruled out that transition be reflexive. The relationship between transitions and fundamental parameters stated in the next chapter will make clear why it might be intuitively meaningful to allow for both possibilities: That some states go into themselves (when the corresponding fundamental parameter is held constant), and that some do not (when the corresponding parameter is changing).

This general notion of transition allows us to define a general notion of relative reversibility. A process $z \Rightarrow z'$ is reversible relatively to this transition whenever $z' \Rightarrow z$ is also the case. Formally:

D3: Rever(z, z') iff: $z \Rightarrow z' \wedge z' \Rightarrow z$.

This is a much broader notion of reversibility than the ordinarily notion—*adiabatic* reversibility—which is, of course, the most interesting one. Not much systematic use can be made of the present general notion of reversibility. However, to express formally some possible situations, it might appear convenient to have that notion at hand.

As for combination, the main requirement is that $\langle \mathscr{T}, \otimes \rangle$ be a commutative semigroup—which is what we would intuitively expect of any interesting physical combination operation (see Duistermaat, [1968], p. 331).

It is required that equilibrium between states be a symmetric and transitive relation, but reflexivity is not required. For some states (namely, non-equilibrium states), it is meaningful to say that they are not in equilibrium with themselves—and, of course, then they will not be in equilibrium with any other state. Therefore, equilibrium is not strictly an equivalence relation.

The last axioms establish some general connections between the fundamental concepts of an operational basis. The first one connects transition with combination; the last one considers the case where a combined state is in equilibrium with a third state.

Some simple and intuitively plausible corollaries can be derived from the former axioms.

T1: If $z \sim z'$, then $z \sim z$ and $z' \sim z'$.

That is, if two states are in equilibrium, they are equilibrium states. This directly follows from $D'2$-(4).

T2: If $z \otimes z' \sim z''$, then $z \sim z'$.

That is, if a compound system is in equilibrium with another system, the components of the compound are also in equilibrium with each other. This directly follows from D2–(4) and D2–(6).

T3: $z \sim z'$ iff $z \otimes z' \sim z \otimes z'$.

That is, if two states are in equilibrium with each other, the compound system is in equilibrium.

Proof: If $z \sim z'$, then, by T1, $z' \sim z'$; hence, we have $z \sim z'$ and $z' \sim z'$; then, by $D2$–(6), $z \otimes z' \sim z'$; and, by T1 again, $z \otimes z' \sim z \otimes z'$. Conversely, if $z \otimes z' \sim z \otimes z'$, by successive application of $D2$–(6), we get first $z \sim z \otimes z'$, and then $z \sim z'$.

T4: If $z \sim z$, then $z \otimes^{n)} \ldots \otimes z \sim z \otimes^{n)} \ldots \otimes z$.

This follows from T3 by iteration in the case $z \sim z$. This corollary might have some interest for compound systems which are supposed to be built out of n replicas of the same system in a given state.

On the whole, not many interesting conclusions can be drawn from the axioms of an operational basis because of their generality. I think this is all right; purely operational concepts should not be considered as very powerful unless they become linked with metrical concepts. The essential physical content of thermodynamics is obtained when both components, the theoretical and the operational, become connected.

VI. THE EQUILIBRIUM THERMODYNAMIC THEORY-FRAME

1. Now, let us explicate the notion of the equilibrium thermodynamic theory-frame as a global structure covering all equilibrium thermodynamic theories. We already have the two basic components at hand: on the one hand, the notion of an equilibrium thermodynamic theory, which is, so to say, the theoretical aspect of

the frame; on the other hand, the notion of an operational basis for thermodynamics, which represents its operational aspect. The frame is the structure resulting from the interconnection of both components.

2. The general intuitive idea behind the interconnection will be this: to every fundamental function of a thermodynamic theory an operational basis will be associated settling the "operational meaning" of that function. So, each fundamental thermodynamic parameter becomes linked with particular forms of transition, combination, and equilibrium.

3. Since every thermodynamic theory includes at least three distinct fundamental parameters, it will be connected with at least three operational bases.

4. Energy and entropy are the two fundamental parameters common to all thermodynamic theories. Their corresponding operational bases will be accordingly "universal" for the thermodynamic theory-frame. In particular, energy and entropy are linked to two well-known kinds of laboratory transitions: transition under *energetic* isolation and transition under *adiabatic* isolation (for short: energetic and adiabatic transitions). These transitions have special formal properties. Energetic transition is always reversible. Adiabatic transition is conditionally connected (as was clearly seen first by Falk and Jung, and so termed by Roberts and Luce). Further, energetic and adiabatic transitions are in a mutual relationship: The first is a special case of the second one. Their relationship to the metric functions energy and entropy is the following one: in an energetic transition, the energy function remains constant; in an adiabatic transition the entropy function cannot decrease. Further, they are in a particular relationship with any kind of (thermodynamically interesting) equilibrium. If a given state is not in this equilibrium, then another state exists which can be reached from the first one by adiabatic transition (sometimes even by energetic transition) and which actually is in equilibrium. Equilibrium thermodynamics excludes from its scope non-equilibrium situations where this does not obtain.

5. All other transitions considered are associated with corresponding metric functions in the same way the energetic transition is associated with the energy function.

6. So-called intensive parameters resulting from the partial deriva-

tion of f^S with respect to one of its arguments are constant when the corresponding equilibrium situation obtains.

7. All fundamental parameters are additive with respect to *any* kind of combination.

8. The energy function satisfies the "energy conservation principle", i.e. whenever the energy is not held constant in a given state, this means that there is another state (possibly a composite one) combined with the first in some way and such that the energy is constant for the compound system of the two states.

9. Entropy and energy satisfy Gibbs' maximum and minimum principles respectively: Whenever the energy of a compound system is held constant in the neighborhood of a certain combination of states, the compound system is in the corresponding equilibrium *iff* the entropy attains a maximum. The analogous statement holds for the energy attaining a minimum.

10. Finally, the *T-theoretical* notion of a (total) equilibrium is linked with the *operational* kinds of equilibrium by stating that a state is in (total) equilibrium when it is in all operational kinds of equilibrium.

11. All these relationships between the theoretical and the operational components of the frame are summarized in the following definition of an "equilibrium thermodynamic theory with operational bases" (*ETTOB*).

D4: $\langle \theta, \mathfrak{B} \rangle \in ETTOB$ iff:

(1) $\theta \in ETT$.

(2) $\mathfrak{B}$ is a set of operational bases and $k(\mathfrak{B}) \geq 3$.

(3) For all $B_i \in \mathfrak{B}$: $\mathcal{T}_i \subseteq \mathcal{T}_\theta$.

(4) There are two distinct B_A, $B_E \in \mathscr{B}$ such that:

(a) $\underset{A}{\Rightarrow}$ is conditionally connected.

(b) $\underset{E}{\hat{\Leftrightarrow}} = \underset{E}{\hat{\Rightarrow}}$

(c) $\underset{E}{\hat{\Rightarrow}} \subset \underset{A}{\hat{\Rightarrow}}$

(d) $z \underset{A}{\Rightarrow} z' \rightarrow S(z) \leq S(z')$

(e) $z \underset{E}{\Rightarrow} z' \rightarrow U(z) = U(z')$

(f) $\neg z \underset{i}{\sim} z \rightarrow \bigvee z_i(z \underset{A}{\Rightarrow} z_i \wedge z_i \underset{i}{\sim} z_i)$

(5) For any fundamental f_i of θ such that $f_i \neq S$, there is a $B_i \in \mathfrak{B}$, $B_i \neq B_A$ such that:

(a) $z \underset{i}{\Rightarrow} z' \rightarrow f_i(z) = f_i(z')$

(b) $D_{f_i} f^S(z) = D_{f_i} f^S(z') \leftrightarrow z \underset{i}{\sim} z'$

(6) For any fundamental f_i of θ and for any $B_j \in \mathfrak{B}$:

$f_i(z \underset{j}{\otimes} z') = f_i(z) + f_i(z')$

(7) $\bigwedge z(DU(z) \neq 0 \rightarrow \bigvee z' \bigvee i(DU(z \underset{i}{\otimes} z') = 0))$

(8) For all z, z', and f_i:

(a) $\Delta U(z \underset{i}{\otimes} z') = 0 \wedge \bigwedge f_j(f_j \neq f_i \wedge f_j \neq U \wedge$
$f_j \neq S \rightarrow \Delta f_j(z) = \Delta f_j(z') = 0) \rightarrow (z \underset{i}{\sim} z' \leftrightarrow \text{rmax}(S/z \underset{i}{\otimes} z'))$

(b) $\Delta S(z \underset{i}{\otimes} z') = 0 \wedge$

$\bigwedge f_j(f_j \neq f_i \wedge f_j \neq U \wedge f_j \neq S \rightarrow \Delta f_j(z) = \Delta f_j(z') = 0)$
$\rightarrow (z \underset{i}{\sim} z' \leftrightarrow \text{rmin}(U/z \underset{i}{\otimes} z'))$

(9) $z \in E$ iff for all $B_i \in \mathfrak{B}$: $z \underset{i}{\sim} z$.

D5: θ belongs to the equilibrium thermodynamic theory-frame iff: There is a $\mathfrak{B}$ such that $\langle \theta, \mathfrak{B} \rangle \in$ *ETTOB*.

From the conditions stated in D4 some general properties common to all theories belonging to the thermodynamic frame can be derived; for example, the positivity of temperature and the so-called principles of Kelvin and Clausius. The detailed proofs can be found in Moulines, [1975], pp. 123–125. They are by no means dependent on the specific properties of the single theory reconstructed there.

VII. A FINAL WORD FOR HISTORIANS

Let me conclude with some general remarks on the potential relevance of this kind of reconstruction to the study of the evolutionary aspects of science. The analysis of thermodynamics offered here is certainly of a synchronic character in its present form, but I think its conceptual categories might be useful for a diachronic analysis of science as well. The main reason is this. If the category of a theory-frame as distinct from single theories is accepted, then a wider perspective in the diachronic study of science, in what we call "the theory of scientific evolution", would be opened. For the term "evolution of scientific theories" would then appear as ambiguous: It may refer either to the study of the evolution of single theories within a given theory-frame or to the study of the evolution of a whole theory-frame as a global structure. In other words, we would have to consider the evolution of theoretical science at two different levels at least. And the examination of the evolution of a theory-frame would have to include not only the evolution of single theories and their inter-connections but also the evolution of what has been termed their "operational aspects". There is a possibility that the evolution of the latter offers a different picture from the one offered by the evolution of theories. All this might still appear as wild speculation, but I think this is a direction which is worthwhile to explore.

Autonomous National University of Mexico

NOTES

[1] As far as I know, the concept of conditional connectedness was first introduced by Roberts and Luce in their (1968). It is particularly relevant to thermodynamical contexts, as we shall see.

[2] The term "belonging" should not be understood in a strict set-theoretic sense here.

[3] The concept of a "physical semantics" and its problems have been discussed with some detail in Moulines and Sneed (1979).

[4] See Moulines, (1975).

[5] For more details, see Moulines, (1975).

[6] Since the formulation of the fundamental law by means of f^S is equivalent to its formulation by means of f^U, in the following only f^S will be considered ("entropy representation").

[7] For the general notion of a constraint, see Sneed, (1971) or Stegmüller, (1976). For its application to thermodynamics, see Moulines, (1975).

[8] See Giles, (1964), Chap. 13.

[9] See Callen, (1960), Chap. 13, §13.1, 13.2.

[10] See Giles, (1964), (1964), Chap. 14.

[11] See Callen, (1960), Chap. 13.

[12] For more details, see Moulines, (1975), pp. 111–112.
[13] See Tisza, (1966), p. 119.
[14] It is remarkable that, although Duistermaat explicitly relies on Giles' work, the last sentences quoted seem to be directed against Giles' conception of combination – even if Duistermaat does not point at the discrepancy explicitly.

REFERENCES

Callen, H. B.: 1960, *Thermodynamics*, New York, 1960.

Duistermaat, J. J.: 1968, 'Energy and entropy as real morphisms for addition and order'. *Synthese* **18**, 327–393.

Falk, G. and Jung, H.: 1959, 'Axiomatik der Thermodynamik,' in S. Flügge (ed.), *Handbuch der Physik*, III/2, 1959, pp. 119–161.

Giles, R.: 1964, *Mathematical Foundations of Thermodynamics*, New York, 1964.

Hornix, W. J.: 1970, 'An axiomatization of classical phenomenological thermodynamics', in E.B. Stuart, B. Gal-Or, A. J. Brainard (eds.). *A Critical Review of Thermodynamics*, Baltimore, 1970, pp. 235–252.

Moulines, C. U.: 1975, 'A logical reconstruction of simple equilibrium thermodynamics', *Erkenntnis* **9**, 101–130.

Moulines, C. U. and Sneed, J. D.: 1979, 'Suppes' philosophy of physics', in R. J. Bogdan (ed.), *Patrick Suppes: A Profile*, Dordrecht, pp. 59–91.

Roberts, F. S. and Luce R. D.: 1968, 'Axiomatic thermodynamics and extensive measurement', *Synthese* **18**, 311–326.

Sneed, J. D.: 1971, *The Logical Structure of Mathematical Physics*, D. Reidel, Dordrecht.

Stegmüller, W.: 1976, *Structures and Dynamics of Theories*, New York.

Tisza, L.: 1966, *Generalized Thermodynamics*, MIT Press, Cambridge, Mass.

SECTION VI

WHAT CAN THE HISTORY AND PHILOSOPHY OF SCIENCE AND PHILOSOPHY DO FOR EACH OTHER?

A Round-Table Discussion

EVANDRO AGAZZI

WHAT HAVE THE HISTORY AND PHILOSOPHY OF SCIENCE TO DO FOR ONE ANOTHER?

History of science has itself a long history, which goes back to at least a couple of centuries ago in the case of some disciplines. Still, it is particularly in the last few decades that it has started to receive a considerable new impulse. But what is more significant is that recent history of science shows a rather new face, if compared with the older one, the novelty being represented by the fact that it shows less and less the features of a "learned" enterprise. Old histories of science had the general character of patient compilations, recording results, discoveries, errors and shortcomings with impartial attention, and framing all that within a series of biographical sketches. The chronological ordering was tacitly supposed to provide by itself the structure of the narration, and this sequential order of justaposition was felt to account for a kind of self-apparent logic of the discourse. Nowadays we are inclined to say that this was no history of science proper, because it lacked the consciousness that science is an historical fact, and this cannot be expressed by the simple consideration that it evolves chronologically. In order to conceive of science as an historical fact, at least two fundamental characteristics must be clearly recognised: (i) that science is one among the factors which make history and, as such, has a complex relationship with the other factors and with the history "produced", which cannot be ignored in making an history of science; (ii) that science is itself somehow "made" by history, it is an historical "product", which cannot be adequately accounted for without considering the more general historical framework in which it is embedded. Let us remark that these considerations are independent (at least to a certain extent) of the discussion concerning the problems of the "internal" or "external" history of science. They simply express the fact that, at last, science too has been included in that "historical way of thinking" that had been customary, *e.g.*, for literature, philosophy, music, and fine arts, but had left science in a kind of limbo of suprahistorical imperturbability.

J. Hintikka, D. Gruender, and E. Agazzi (eds.), Pisa Conference Proceedings, Vol. II, **241–248.**

It is this new way of perceiving science that not only has led to a new style in the history of science, but has also determined a number of mutual relationships between the history and the philosophy of science which we shall now briefly examine.

Considering philosophy of science in itself, one must say that it is, in a way, an effort for characterising "what science is" or, if one prefers, that it is a "theory" about science in a sense which is rather similar to that in which every specific science is a "theory" about its proper domain of inquiry. It follows that philosophy of science cannot be constructed arbitrarily: it does not express the views of this or of that philosopher about what science ought to be. In other words, it has to satisfy some kind of "empirical requirement". This has always been more or less understood, but it was not very clear when and where such an empirical testing of the different positions upheld in philosophy of science was to occur. As a matter of fact, a tacit admission was that all the discourse of philosophy of science is necessarily simplified and schematic, so that only some idealized patterns of actual science might be taken into account in it. It followed that only some very generic and rather insignificant patterns of present science (especially of mathematics and physics) served as the implicit empirical reference of most constructions in "general" philosophy of science. On the other hand, "special" inquiries in the philosophy of science used to be concerned only with some very technical topics of present science as well (like, again, certain foundational questions in mathematics or some central points in relativity theory or quantum mechanics). In both cases, it seemed a matter of course that science is *present* science, and that the science of the preceding times was only an imperfect anticipation of it. Rather insignificant exceptions were constituted by the generic mention of some stereotyped events in the history of science, like the Copernican revolution or the construction of Newtonian mechanics, which were hinted at with no more precision than the vague reference permitted by some souvenirs of school education. Now, on the contrary, it has become more and more clear that history of science provides the more natural and fruitful source of examples and counterexamples for discussing controversial issues in the philosophy of science, as science is not a generic and idealized entity, but a concretely existing process with a rather long history, in which different theories, different instances of conflicts between theories, of actual ways of behaving, of theoretical and empirical aspects, *etc.* are to be seen "at work", and this is an enormous contribution that history of science makes to a more rigorous and "testable" way of doing philosophy of science.

But there is still more than that. If science has an historical nature, as we have pointed out before, it follows that no adequate philosophy of science can overlook this fact, otherwise it could not be an inquiry proportioned with its own object. On the other hand, history cannot be invented or described *a priori*, so that one has to resort to some actually and correctly performed research in the field of history of science, if one wants to take into consideration the historical nature of science in a not purely postulational way. This means that not only ought philosophers of science to be more familiar than they often are with the history of science, but they should, at least sometimes, do some research in this field and not simply be satisfied with second hand presentations. For nothing can replace the lesson one gets from real and direct contact with science as a basis for a convincing philosophy of science; but this contact can only come either from some familiarity with working, present science, or from an effective immersion in some concrete moment of past science, via first hand historical research (better if both experiences may be put together).

There is, moreover, sufficient evidence that the perception of the historical nature of science has directly affected philosophy of science as such. It is enough to consider the typical shifting in the standard configuration of philosophy of science that has occurred in the last two decades. One could say that we have passed from a "structural" to a "dynamical" consideration of science and the salient topics of philosophy of science reflect this transition quite faithfully. Indeed, while the most investigated problems in standard philosophy of science used to be, let say until the late fifties, those of concept formation, theory construction, validation of theories, inductive support, empirical versus theoretical components of theories, *etc.*, the most debated questions in the last years have been connected with problems like those of theory change, scientific revolutions, or progress in science: that is to say, with problems which were only seemingly included in older topics, such as that of the transition from one theory to another as a consequence of clashes with empirical data.

In order to make this difference clear, one might say that the old way of considering the transition from one theory to another consisted in considering it as a problem of the "logic of science". In other words, the new fact which suddenly emerged and showed incompatibility with an existing theory, was meant to produce a logical conflict which was supposed to be eliminated by the creation of a (more or less) new theory. As for the relationships between this new theory and the old one, the standard empiricist view was that the new had to include the old as a sub-theory (ensuring in such a way a quite easy cumulative progress in science). The more sophisticated Popperian view was that the two theories were incompatible, but the new had somehow to explain, from its own point of view, all the facts the old theory was able to cope with, and some other facts in addition. If one considers the positions of Lakatos, Kuhn and Feyerabend, one can say that they are a kind of extension and refinement of the Popperian attitude (and in a certain sense they actually are). But this interpretation misses the overall important point that logical requirements play a very small role in the description and justification of theory change given by these authors (especially by the last two of them). They even show that, in many cases, logical requirements are neglected or patently violated in the actual transitions we find in history. In such a way, the key for accounting for theory change becomes a quite different one: no more a logical, but an historical account is given, and the categories involved are those familiar to historians of institutions and of ideas, rather than to philosophers of science. Sociology of knowledge, even psychology, are called into play in order to interpret theory change, in a way which intends to be "rational", although not strictly "logical". This on the other hand does not prevent structuralistically minded people from trying some interesting "logical reconstructions" of these patterns, as it happens with the well known Sneed–Stegmüller interpretation of the Kuhnian ideas, but this does not modify the specific character they show in comparison with the older views.

There is even the possibility that the effective attention paid to the actual history of science as a consequence of the new way of conceiving science, may lead to major modifications also in some patterns of the best established "structuralistic" conceptions. A good example of this kind seems to me the drastic limitation of the whole Popperian way of conceiving science which has come from Kuhn's analysis of science. As is well known, Popper identifies scientific inquiry or even scientific mentality with a perpetual effort of putting our own theories (or "conjectures") to the most severe tests in order to falsify them. This picture is certainly fascinating, and everybody has probably experienced the strength of this presentation when he first met it. But Kuhn has clearly shown that this is an idealised behaviour that belongs to the ethics of the pure scientist, rather than a faithful description of what science really is. For by far the most frequent and long periods of "normal" science are those in which specialists work under the unquestioned authority of some "paradigms", aiming at solving puzzles by means of the theoretical and methodical tools provided by them, and never having it in mind to challenge them and to see whether they can be dismissed. Only in the comparatively rare and short periods of "extraordinary science" does the Popperian schema apply with reasonable approximation.

Needless to say, this historically dimensioned way of conceiving science has a lot to say, also, about the question of the cognitive value, the ontological commitment, and the relation with truth that scientific knowledge can admit of. In general, the whole problem of "scientific objectivity" receives an illuminating reformulation when it is revisited without the crude presuppositions of the existence of "pure facts" from one side, or of "pure conventions", from the other, which do not leave any room between a naive dogmaticism and a gratuitous phenomenalistic conventionalism. The idea of the "historical determinateness" of the scientific objects (which the present author has tried to develop elsewhere) may give an equilibrated solution to several problems in this context.

We have spoken of the impact of the history of science on the philosophy of science thus far. We now want to say something on the reverse influence as well.

A first quite obvious consideration is that, as nobody can make the history of what one does not know how to identify, at least some idea of science must be tacitly presupposed by every history of science. At worst this idea is given by the current common sense notion of what the sciences are, but it is quite clear that a better understanding of the nature of science, such as can be given by a philosophical reflection, is destined to substantially improve the quality of the history of science we can do. Evidence of this fact can be given by the already mentioned rather insignificant character of several ancient histories of science, which were written having in mind only the generic pattern of the sciences present at that moment, without any illumination of a philosophical nature that could make the presentation more problematized, articulated and illuminating. And one should not believe this to be simply a question of elegance, of literary performance, or psychological attractiveness. The real point at issue is that, without a certain comprehension of the intrinsic nature of science, the history of science cannot find its specificity and seriously risks becoming a bad hybrid of other types of history. Let us only think of the anecdotical model applied to the history of science, which has given rise to so many uninteresting productions because of having applied a literary model to a subject matter which was rather refractary to it. The same can be repeated for the nude catalogue of discoveries and errors that we find in other cases, and which comes from the fact of not having had any idea of how to organise that material in a "narration" which could be specific for science. The much more mature philosophical consciousness of our historians of science accounts for the better standard and the more incisive interest their works possess in comparison with the past.

A somehow more intrinsic witness of the impact of philosophical questions upon the history of science is given by the well known controversy about the "internal" or "external" history of science. As has already been stressed, this is a question deeply rooted in the general conception of science (indeed, authors like Lakatos have devoted a major part of their work in the philosophy of science to the discussion of this problem or of questions directly related to it). The main point, of course, is not to say whether a good history of science should be a purely (or prevalently) internal or a more clearly external one: it would not be difficult to recognize that both aspects are to be present and that some intermediate position should be adopted. But then the difficult task is that of establishing which facts belongs to the "internal" and which to the "external" history of science: an empiricist or even Popperian way of conceiving science would probably exclude from the internal history much that other conceptions are likely to include, and this would immediately give rise to histories of science of quite a different character.

There is a warning that is worth mentioning in connection with this influence that philosophy of science exercises on history of science: the risk is that somebody could be led to made an *ad hoc* history of science in order to support some particular theses in the philosophy of science. Such a danger has a quite natural theoretic explanation: if it is true that philosophy of science is a kind of "theory" about science which is being put to the test by means of evidence produced by historical investigation, it is only all too natural that once somebody has proposed some hypothesis in philosophy of science, he tries to find in history of science some good example supporting it. Unfortunately, while empirical tests in the case of exact natural sciences are rather difficult to manipulate or to "interpret" in a more or less arbitrary way, in order to let them cope with a favorite hypothesis, in the case of history (as well as of other "human sciences") the freedom of the "interpretations" is much wider and one is always inclined to consider as "irrelevant", or negligible, facts or details which do not fit in well with one's ideas, and to overestimate other facts or details which are in agreement with them. Beside this purely theoretical explanation, there are, unfortunely already, several examples of a partisan history of science, done with the implicit or explicit purpose of supporting certain particular conceptions of science. It is this way of doing history of science that is about to discredit history of science itself, or at least to make people suspicious about the usefulness of admitting inferences of philosophical outlooks in the history of science. In such a way, many of the positive results which were obtained as a consequence of this interplay risk being overthrown by the misuse of some people.

What ought we to do about that? One thing should be clear at any rate: that nobody can make any history of sciences without having an implicit or explicit picture of science in his mind, and, such being the situation, it is much better to have an explicit and conscious one than an implicit and unconscious one. This explains why it would be naive to plead for an uncommitted or agnostic history of science, which would only reduce us again to the old positivist model of chronological records of unilluminating and arbitrarily selected facts. A selection has to be made in any case, as history (in all domains) cannot record all the events, but needs to limit the narration to the relevant or most significant ones. But now one sees that *criteria* of relevance or significance come immediately into play, and these are the expression of the different conceptions one has, *e.g.* in our case, of science.

Nevertheless, this does not necessarily imply distortions or misinterpretations of the history, and here we must vindicate a specific "professionalism" of the historian of science, who has to be a true historian first of all. This professionalism, which induces the habit of respecting documents, of being extremely careful in generalizations or in hasty conclusions, which is suspicious about too enthusiastic syntheses, which is expecially conscious of the fact that real events can have a quite bizarre course and not fit in well enough with our preconstituted schematizations, can be the best antidote against the *ad hoc* way of doing history of science we have mentioned before. As a matter of fact, it is quite easy to recognize, in the examples of such pieces of bad historical reconstructions we already have at our disposal now, that specific competence in the field of the particular science investigated, adequate knowldedge of the general history of the period involved, and sufficient professional competence and training with historiographical tools and methods are far from being in the possession of their authors.

We can, therefore, draw a kind of general conclusion from all our discourse. We might say, paraphrasing Kant, that history of science without philosophy of science risks being blind and that philosophy of science without history of science risks being empty. But we must also say that history of science supported by a dogmatic and pretentious philosophy of science risks being doubly blind, while a philosophy of science supported by a partisan history of science risks being both blind and empty at the same time. This is to express the delicacy that the mutual relationships between history and philosophy of science intrinsically imply: like every serious and constructive cooperation, they need much effort of reciprocal comprehension and understanding in order to be instrumental to one another without losing their proper connotations, in order to reach results that the restriction to one field only would never have permitted us to reach.

L. A. MARKOVA

A COMMENT ON E. AGAZZI, 'WHAT HAVE THE HISTORY AND PHILOSOPHY OF SCIENCE TO DO FOR ONE ANOTHER'

Professor Agazzi proceeds from the fact of a serious crisis in the positivistic philosophy of science which is, in essence, anti-historical, and brings forward the post-positivistic trends in philosophical thought associated with K. Popper, I. Lakatos, T. Kuhn, P. Feyerabend, and others. It has been rightfully noted that the interest revealed by contemporary philosophers of science in the growth and development of scientific knowledge makes cooperation between historians and philosophers more natural and fruitful.

An analysis of the crisis in the philosophy of science over the last two or three decades provides an opportunity for in-depth penetration into the specific features characteristic of the positivistically-minded philosophy of science of the past, and the essence of its connections with the history of science. According to Professor Agazzi, the anti-historical nature of positivistic philosophy found its expression in the treatment of science's past as a mere pre-history of the present; it was assumed that everything of value brought to life by the past enters the modern structure of scientific knowledge. The past interests us as a preparatory phase for the present, as its anticipation. Present day scientific theory acts as the starting point in writing history. When taken by the philosophers of science, such a stand hindered their cooperation with historians, whereas writings in history generally lacked conceptuality; they were of a descriptive nature and did not go much beyond mere narration of events in their chronological order.

On the whole, one can accept the features of positivistic philosophy that Professor Agazzi mentions in his report. However, when Professor Agazzi defines the place of the positivistic philosophy of science in history, thus himself posing as a historian of philosophy and the historiography of science, his view reveals some aspects severely criticized by him as being characteristic of the positivistic historiography of science. The fact is that the positivistic philosophy of science itself needs a historical approach in both its evaluation and criticism. The very antihistoricism of positivism gave rise to a certain type of historical research, and philosophical concepts in history. The

J. Hintikka, D. Gruender, and E. Agazzi (eds.), Pisa Conference Proceedings, Vol. II, 249–251.

treatment of these notions as inferior ones – as concepts which no more than anticipate the outlook of contemporary philosophers of science – is contrary to history in the same sense in which the past history of science was treated by the positivists. If contemporary historians and philosophers of science are to depart from the positivistic approach to past science, they have to abandon the positivistic viewpoint in analyzing past philosophy and the historiography of science.

The positivistic conception of the history of science was quite complex and problematic. Although past science was depicted as antecedent history and a prerequisite of the present, it meant (and both historians and philosophers of science frequently wrote about it) that each new fundamental discovery in science lead to the rewriting of its history anew, in conformity with the new theory of natural science. It was as if two revolutions were concurrently underway: one in the natural sciences, and the other in history; and the historian was faced with a very difficult task involving the solution of quite a number of problems, not only in history but also in logic and the natural sciences. The preparation of a paper in history, even a descriptive one where the major task consists in the chronological ordering of past events, makes it pertinent that the historian should be well-versed in contemporary theory and the logic of natural science, according to which he will select from the past and arrange in a single logical sequence only those events which have prepared the emergence of his contemporary knowledge.

One cannot deny that the positivistic philosophy of science contributed to the shaping of a definite type of philosophy of history which has been basic to most works in the history of science for many decades. Today we cannot feel satisfied with the main features of this philosophy (above all, its cumulativism, the approach to progress as a gradual and continuous process, and its rejection of the idea of revolutions in science). But this does not mean that one can ignore its existence as an historical event, and it does not permit us to avoid a profound understanding of those theoretical difficulties which resulted from the development of the positivistic conception. The awareness of these obstacles will make it possible to grasp the present situation in philosophy and the history of science.

The participants of the "Round Table" discussion, including Professor Agazzi, emphasized the historical nature of the contemporary philosophy of science which has replaced positivism. Nowadays, contrary to the positivists, attention is concentrated on discontinuities in the progress of science, on revolutions in science. In the form taken by the corresponding research, it inevitably encounters hindrances previously observed in general history, for instance, by Spengler and Toynbee: the past is more than mere anticipation of the present, it is unique and original. But then, how is communication with this past possible at all – how can the transition from the past to the present generally be achieved? Probably, this problem is the most crucial one in the contemporary philosophy of science, and it is also the most difficult to solve. It is in this context that there emerge questions about the nature of rationality, the logical and the socio-psychological, the personal and the social, the subjective and the objectively necessary, and so on.

It is rather peculiar that nowadays the most difficult problems are frequently those which were solved quite easily within the framework of the positivistic conception, such as the problem of revolutions in science. Cumulativistically-minded historians are sure to have recognized the phenomenological presence of great, fundamental discoveries in the history of the natural sciences and they classified these as revolutions in science. But, they believed, in order to be understood, these revolutions are to be incorporated into a continuous series of developments. This operation was actually performed indiscriminately in the above-mentioned way: every achievement of science in the past was treated as part of the antecedent history of contemporary scientific knowledge. At present, the problem of revolutions in science is approached differently. They are supposed to be understood as a kind of discontinuity within the continuous development sequence, as points at which fundamentally new knowledge emerges; however, this knowledge does not cover all the previous progress, and admits the self-sustained existence of previous knowledge. Thus, history is guided by the principle of correspondence, or the principle of complementarity taking shape within the framework of physical knowledge. By referring to the above example of dissimilar approaches to the problem of revolutions in science, I should like to emphasize, yet again, that the positivistic conception of history was something more than the antecedent history of our modern subtle and ingenious judgements: it was a comprehensive philosophy of history justified historically and socially.

ROBERT E. BUTTS

METHODOLOGY AND THE FUNCTIONAL IDENTITY OF SCIENCE AND PHILOSOPHY*

I

Scholars put the question expressed by the title of this roundtable discussion with some urgency nowadays. It is often as if the questioner thinks that the very continuation of the two specializations depended upon an answer to the question *now*. Things do not seem to me to require the sense of urgency, any more than they are made better by one of two relatively-complacent attitudes: "Well, the importance of history to philosophers of science has now been demonstrated, let us turn to questions of what, if anything, philosophy has to offer the historian of science"; or, "Well, we know that all history of science is infested with philosophical considerations, let us turn to the question of whether or not historians have anything relevant to say to philosophers of science." I think it is not often enough appreciated that both history and philosophy of science, as special research and teaching disciplines, are relatively new. After all, it is only in our Century that history of science has become a fully mature and self-conscious technical discipline, just as it is only in our Century that philosophy of science, armed with more powerful technology in the form of logic and techniques of analysis, has become, rightly or wrongly, a separable aspect of philosophy.

Earlier in this Century both historians and philosophers of science walked narrow paths. Historical work tended to be locked within the confines of expository or narrative accumulation of the details of the past of science, and to building the bibliographical background required to develop this form of history. Philosophers of science, or at least those who turned away from the tradition of combined interest in both history and philosophy of science furthered by the work of scholars like Whewell, Duhem and Peirce, created an almost fraternal model of science as psychology-free, history-free, value-free; free, indeed, of almost everything but the tools of logic and a maturing semantics. How could it have been otherwise?

J. Hintikka, D. Gruender, and E. Agazzi (eds.), Pisa Conference Proceedings, Vol. II, 253–270.

Faithful to the ordinary demands of customary historiography, the work of early 20th Century historians of science was as it should be: the data needed to be supplied before more enriched speculations about the nature of science could be undertaken. The same kind of professional scholarly constraints (constraints that need not be associated with attendant paranoia) acted upon philosophers of science, and, within the limits of those constraints, the new technical philosophers did what first needed to be done; namely, to attend in earnest to the genuinely philosophical aspects of science: argument and analysis. Historians quite correctly began their inquiry with an attempt to narrate the details of purported contributions to knowledge made by scientists in the past. Philosophers quite rightly sought to understand scientific practice as a complex form of reasoning requiring, before anything else of use could be accomplished, a technical analysis of methodologies and forms of scientific inference. That some of the more breathtaking joys of both history and philosophy were left out of consideration by these earlier researchers was thus necessary for *methodological* reasons. The very nature of their work rendered most historians and philosophers incapable of realizing this cardinal point. Inevitably, the very myopia engendered by the constraints of their own work led each group to begin to mistake the motives of the other.

Thus was created a tension that is with us today, although, hopefully, with diminished effect. It is often the case that different (*different*, not necessarily *competing*) technologies will result in different and, at least, thought-to-be competing ideologies. I suggest that it is the reproduction of this syndrome in the work of early 20th Century historians and philosophers of science that has brought us to the question raised by the title of this session. Historians, in one sense at least rightly confined by their technology and methodology, regarded philosophers as evangelists trumpeting a heretical picture of science as part of a larger world view. Philosophers, in one sense at least rightly questioning the methodological relevance of historical detail to the task of explicating the forms of scientific inference, accused the historians of foisting upon us an ideology of science based upon the commission of the simple logical fallacy of taking the genesis of an idea to be relevant in establishing the validity of that idea. No sensitive scholar in either history or philosophy of science can today be fooled by this syndrome. With hindsight it is just too easy to discern that differences are not necessarily competitions, and that we should learn from this that impaired vision resulting from methodological constraints is, in any and all fields of rational endeavor, subject to the *general* methodological imperative that we should investigate the limits of our own ideology before we accuse others of creating grief through the questionable introduction of their ideologies.

It is precisely in this connection that the work of scholars like Koyré, Burtt, and Crombie is of such extraordinary importance. These thinkers, well versed in history and in philosophy, first pointed up the poverty of applying either the technology of history or the technology of philosophy *exclusively* in an effort to reconstruct a view of the nature of science. Existing, sometimes, quite apart from these methodologies are the richer aspects of the cultural, social, and metaphysical background of science. We need also constantly to remind ourselves of those in both history and philosophy who pursued their work in full awareness of the methodological imperative mentioned above. For example, it was the historian Kuhn who located what might be the limits of history in understanding science. His work on so-called scientific revolutions has repeatedly emphasized the socio-psychological factors that limit science, and in the bargain also limit any narrowly historical or philosophical treatment of the nature of science. On the side of the philosophers, many of those who would be listed as positivists or logical empiricists have, in the investigation of their analytic models, discovered and reported limitations of their models, or have shown willingness to expand their analyses beyond the cramped confines dictated by an earlier application of logic and other analytic tools. There are many examples; for now, think of Goodman and Quine and the enlarged context of pragmatism; or of Hempel and Rudner and the enriched understanding of scientific explanation.

Nevertheless, and this in spite of the happy developments I have mentioned, the parallel progress of history of science and philosophy of science today is not overwhelmingly harmonious. I must here admit to a certain bias, albeit a justifiable one: it is the philosophers and philosophically minded historians who have begun to bridge the gap between the two fields, and it seems to me that this is as it must be. It is up to the philosophers to advance the cause of cooperation. There are, and perhaps I will be excused for pointing this out, some simply *historical* considerations that weigh heavily in favour of this posture.

As a professionally organized group, historians of science outrank philosophers of science in the 20th century sense. This has a crucial bearing on the fact that philosophically informed and interested historians of science are really rather hard to find. At least this seems to have been the experience of members of the Joint International Commission on History and Philosophy of Science during the past six or so years. Perhaps this helps to explain why in recent years historical work in logic, methodology and philosophy of science has displayed enormous growth (as evidenced by the number of contributions in these areas made during recent congresses of this Division of the International Union of History and Philosophy of Science); whereas historical accommodations of philosophical interests in science appear to have grown very little (as evidenced by the very skimpy offerings, if any at all, made by philosophers or philosophically oriented historians during recent congresses of the historical Division of the International Union of History and Philosophy of Science). I want now to put forward what may seem a very daring thesis in an effort to provide a framework for understanding this curious situation, and to help in furthering the cause of future cooperation between the history and philosophy of science.

II

The thesis I want to defend, for, indeed, I think it must be defended, would at one time in history have appeared to be a platitude: *there is no interesting and substantial difference between philosophy and science*; *at least no difference that can mark off the two enterprises as different in kind.* I will call this the thesis of the *functional identity* of science and philosophy. Platitude or not, the message has been forgotten by many recent philosophers, and seems entirely to have escaped the notice of most historians of science. Just consider the following suggestions. What is being done in the intellectual activity of a theoretical physicist is no more, and no less, *speculative* than, say, Wilfrid Sellars' visionary and profound reconstruction of parts of what he calls the "scientific image". Both activities are constrained by a tightly structured rational syntax; both "solve" problems by massive efforts at analysis and systematization; both are controlled by prevailing canons of parsimony; and, *perhaps most importantly*, the *worth* of both activities is determined neither by direct and unambiguous appeal to "facts", nor by considerations of "application" or pragmatic fit, let alone by considerations of direct usefulness. If considerations of this sort were uppermost in the minds of referees of funded research projects, no theoretical work in science, and no philosophical work, would ever be funded – would ever be paid for out of hard coin.

At what is often taken to be another extreme, the importance of the scientific activity is measured by the expectation that results of theory will, at least some of them, be amenable to experiential confrontation. The measure of scientific acceptability is thus in a deep sense epistemological – scientific truths are acceptable just in so far as they reflect accurate experimental or observational results. It is often thought that philosophical speculation or analysis lacks this experiential "hardness". It is argued that what science possesses by way of experiential control is what is precisely lacking in tests of philosophical results. Nothing could be further from the truth. Unless we are to accept the most cynical psychoanalytical view of philosophies as generated *solely* by the unique psychological needs and aspirations of each individual philosopher, I think we must admit that, in some indirect way, philosophies, just like bits of scientific theorizing, must be confronted with some test of experience, where experience, although in some sense theorized, is so minimally theoretical as to constitute something like a publicly available datum.

Of course what we take this or that datum to "mean", what we take to be the "beings" lying behind the data (what are sometimes called our ontological commitments) may vary from individual to individual, culture to culture. All that I am claiming is that the bridging of the cultural or individual gaps does not require any more in the way of theory than is ordinarily needed to "translate" one's experiences into the "language" of others. It must thus be said on Aristotle's behalf that, in the sense of public experience I am suggesting, he, along with any good philosopher, would find a piece of science and a piece of philosophy abhorrent for exactly the same reason if indeed it was in no way amenable to any kind of confrontation with experience.[1] It is important always to guard a fairly hard distinction between ontological (or conceptual) commitments, and truth claims. One is entitled to a private ontology (as in dreams and constructed fantasies); one is never entitled to a private truth. Gnosticism may work as an ontoloical motivation; it can never work as an epistemologically justified methodology.

Now, if the thesis concerning the historical continuity of speculation, methodology, and the role of experience in science/philosophy and the epistemological barrenness of any attempt to separate the three is correct, as I think it is, then certain interesting consequences follow that touch vitally upon the question embodied in the title of this session. For one thing, if science and philosophy are essentially the same kind of activity, then we can determine at once what it is that historians of science can do for philosophy. In my view, the history of philosophy is simply one part of the history of science (or the history of science is simply one part of the history of philosophy). Thus historians of science have an obligation to work in the history of philosophy; it is part of their field.

Are we onto something important here? There have been very few important philosophers who did not know the history of their discipline. Today, mastery of the history of philosophy by well-trained philosophers is taken for granted. Conceptualization of experience has a history; theoretical explanation has a history. But what about a corresponding cooperative element in the training of historians of science? Has the mastery of the structure of conceptualization, a grasp of the logic of scientific inference, a familiarity with the canons of experiential methodology, been left out – perhaps ignored? Discoveries are not understood if all that we do with them is list them, date them, woodenly correlate them with a man (or woman) and a time. Discoveries are conceptualized expectations; both a logic and a methodology predate each one. Is the historiography of the historian of science too restricted? That, too, is a philosophical/scientific question.

What the historian of science, then, can do for the philosopher of science is exactly the same as what he can do for the scientist – in the best of cases, he can display the rich and complex conceptual, methodological, and experiential components of this or that piece of theory, this or that discovery, this or that failure in application (or success in application – I am thinking now of the history of technology). One would hope also that such historical work will be deeply evaluative, for one of the things we need to know about science is why, for example, Galilean ways of viewing the world (once these ways are clarified by the application of more refined kinds of mathematics than Galileo himself had to work with) are *better than – are improvements upon* – Aristotelian modes of structuring experience. If we cannot arrive at something like answers to this epistemological/methodological question, there would seem to be little point in regarding the history of science as helpful in understanding science. It is, after all, scientific *success* that we want to emulate, and success is a value (as, of course, is failure). But this gives us an incomplete answer to the question posed by the title of this discussion. We must turn to the question of what it is that the philosopher can do for the historian. This question, as I have already suggested, has a clear logical priority because until it is answered clearly, the historian, I am afraid, can give nothing – nothing, that is, unless one is prepared to settle for a history of science as a history of test tubes and triumphs of discovery, just as some are prepared to accept standard history as a history of beds and battles.

I should say at once that I also think it correct that the philosopher of science can give nothing unless he understands the philosophical/scientific activity as having a history. Indeed, one could argue that the very meaning of his problems is in large part supplied by that history. Can one genuinely be said to understand the quest for charming quarks unless one also understands the kinds of things Thales and Anaximander were seeking to know? Certainly *not* Socrates, who was a social prophet and an unemployed, if very sophisticated, lawyer; it was Plato who made him into a philosopher/scientist. Or, can one genuinely be said to understand the explication of the concept of analyticity without a knowledge of the sources of this concept in mathematics and methodology? Philosophers who diminish history for the sake of incubating logic betray themselves. If what I am saying begins to generate a circle, begins to sound like: one cannot do history of science properly without a knowledge of philosophy/science, and one cannot do philosophy/science properly without a knowledge of history (so that although there is a logical primacy to the question 'what can the philosopher/scientist do for the historian of philosophy/science?' the two questions, in a larger, almost organic context both presuppose answers in the same time; de facto answers, if you like), well then isn't that exactly what we had hoped for – this happy circularity – when we embarked upon the discussion occupying us this past week?

History, just like philosophy in some of its concerns, such as logical analysis and explication of concepts, is without a fixed subject matter, and hence without a standard *general* methodology. Illuminating history, *explanatory* history, therefore presupposes frameworks of interpretation, goal oriented applications, and adaptable methodologies. Thus history, and here it is in positive respects much like the empirically theoretical and methodological aspects of philosophy/science, is not self-critical – its frameworks, applications, and methodologies are imported, extrahistorical. There is of course no harm in this, provided that history at some point makes up the balance of payments, and it cannot do this simply by producing more history on the basis of previously imported conceptual and technological considerations. There comes a point at which history ignores its philosophical presuppositions at its peril, for that is tantamount to losing focus, and consequently to losing interest, except for those with the most narrow historical concerns.

No face is lost in this effort at enlightened self-consciousness, for the best strategic models providing frameworks, applications, and methods for history are both philosophically committed and historically biased. I will not here offer argument for this contention, nor will I pretend at this point to be able to offer the historian of philosophy/science a pat presuppositional conceptual background for his work. What I will do is sketch briefly a model for understanding science in a historical context, a model in which various longstanding philosophical problems may someday be seen to be dissipated with increased attention to science as a process and an enterprise that grows and changes over time. I would hope that the details of the model, once worked out, would illuminate the nature of science; for the moment, I offer the model as the basis for conceptual cooperation between historians and philosophers.

III

My model embraces three aspects of the scientific enterprise; for the sake of brevity I will refer to them as the institutional, the methodological, and the epistemological aspects. I do not mean to suggest that science operates in three separate and distinct contexts. Quite to the contrary, the three aspects of science are interconnected, developing, and changing functions of the enterprise; they will be separated here for the sake of the analysis only.[2]

I regard it as a datum that all intellectual quests, including those of the scientist, are undertaken for some purpose. One endeavors to solve an intellectual problem for the satisfaction of some goal that exists independently of the activity of finding the solution. This is true, notice, even if the aim is the satisfaction of curiosity (one may be tempted here to say "mere curiosity", as one often does; but I think the satisfaction of curiosity is a basic human need requiring no external justification and no defence). The interesting thing to note is that examination of the work of individual scientists does not always, or even often, reveal the aims for the sake of the satisfaction of which the work was undertaken. This is partly because the attempt to satisfy a need often becomes automatic and is, in any case, self-justifying. It is also because the aims of the work do not appear to be essentially involved in the work itself. It is for this reason, I believe, that persons who do highly theoretical and speculative work, work that at least appears to be marvellously abstract and distant from more ordinary human concerns, recoil from the query: "But *why* are you interested in *that* question?", or, "What *good* is a concern for that subject?" The discomfort generated by such questions is at least in part a function of the fact that the questions do not have answers *that are truly required to be given.* After all, the questions are conceptually analogous to: "Why do you eat?", and "Why do you seek to be warm?"; the answers being, obviously: "Because I am hungry", and "Because I am cold". But one is right to be offended by having to give such answers; the offense, after all, is that the question ought not to have been raised in the first place.

As the aims of the work become less prominent – and this seems true no matter what those aims were – the work itself abstracts itself from the continuum of ongoing human concerns, and soon becomes the kind of activity that can be ascertained to be "professional", that is, that can be fraternally protected within the comfortable confines of this or that institution (or discipline, as Rescher would have it). This institutionalization of science results in a network of professionally approved, fraternally accepted ways of engaging in the work of the scientific enterprise. This approval and acceptance, however, is purchased at a high price, for now the institution, operating free of its initial goals, takes on an inevitably political cast: the scientific institution must compete for a place amongst other institutions. Once the preserve of both the wealthy and the ne'er-do-well (the aboriginal amateurs), in its middle period financially advanced by patrons and brilliant dilettantes, the scientific enterprise is now a socio-political competitor for the goods of society, the need for a governmental science policy having become both a moral and an economic necessity. And, given the apparent narrowness of its interests and the esoteric and mysterious nature of its work (remember that the initial aims are now forgotten in the abstraction of the enterprise from those aims), science is a poor institutional competitor. Thus the scientific institution seeks "justification" in engineering applications and in fostering technological advances. In this respect, science betrays its original aims. For politics, as we know, removes the fun from the original activity, and becomes an end in itself. Perhaps this is why politics itself seeks to become a game? After all, one either wins or loses in politics.

It ought not to be thought, however, that the institutionalization of science is entirely disadvantageous. For both the historian and the philosopher of science, the abstraction of science from its initial goals and aspirations has important advantages. For one thing, neither the historian nor the philosopher need seek excuses for the fact that in studying science he deals with an abstraction, for it is not the historian and not the philosopher who creates this abstraction; rather, he simply finds it among the givens of his data. Consequently, the minute investigation of the details of a scientific discovery or the analysis of scientific systems as systems of explanation and confirmation – investigation or analysis uniquely presupposing the availability of detemporalized slices of the scientific enterprise in abstraction – is fully justified by the "facts". This abstract nature of disciplinary science supplies the reason why R.B. Braithwaite could assert in such a blasé and uncommitted manner in his introductory lecture in the undergraduate course in philosophy of science that what he was dealing with is "dead science". It also explains, I think, why the historical curiosity of dating manuscripts (an activity curiously and paradoxically bent upon establishing the *atemporality* of the scientific enterprise), and the absolutely colourless examination of the details of manuscripts (as, for example, in the philosophically vacuous investigations of Stillman Drake), can be so consumingly fascinating as ends in themselves. The point is that all such studies are together justified by the disciplinary or institutional character of research.

These various advantages notwithstanding, the most important benefit of the institutionalization of science comes forward when the detemporalized slices of the activity, historically investigated and philosophically analysed, are put back together (or, better, we attempt to put them back together). The reassembly of these dated and conceptualized slices requires a cipher. It is not enough to say – what may be true – that the slices are temporally related, that they follow one another in time. The slices following one another in time are like Hume's sense impressions following one another in time: they are related to one another in no *meaningful* way by being seen to be historically connected. Each detemporalized slice, authenticated and properly dated, comprehended in its logical structures of explanation and confirmation, bears a relationship to other such slices only on the basis of the imposition of the cipher that allows us to put the pieces of the puzzle together. The solving of a puzzle takes place through intervals of time, but the solution is not given by this activity; it is, so to speak, *there all the time.* I would suggest that this cipher is intimately related to the *success* of the science represented in some of the slices or bits of the puzzle, and that the question of success or failure in science drives us immediately beyond questions of the *content* of any or all of the slices into a consideration of the *methods and techniques* (*technologies*) employed in just those cases that we take to be the best exemplars of scientific work.

Larry Laudan has argued persuasively that it is not the philosophers, but the scientists themselves, who first and foremost articulated the most interesting and successful methodologies.[3] I am not inclined to accept this argument without modification to accommodate my view of the identification of philosophy and science, but in dealing with at least some of the abstractions presented by disciplinary science, the point of his argument is entirely correct. In any case, Laudan and I are as one in thinking that the study of methodology is the most promising bridge for relating the efforts of historians and philosophers. To cite a case in which I have myself been at least marginally involved, look at the recent work on Galileo by both historians and philosophers of science. Studies of the detemporalized slices continue: not all of the authentication and dating of manuscripts has been accomplished, the logic has not been fully understood, but enough of the science of Galileo is accessible to provide new insights into how the various Galileo slices fit together.

The insights, of course, are insights into Galileo's methods. One would hope that it is not accidental that recent work by both historians and philosophers has converged upon an attempted reconstruction of Galilean methodology. There are many reasons for this convergence to be occurring now, nevertheless, one hopes that all the work being done is based upon accepting that Galileo went about his business in – if not the right way – a way holding out great promise of success in certifying truth claims about the world (methodology is not philosophically neutral; the epistemological presuppositions and consequences are there for all to see). Such acceptance of Galileo's work is by no means gratuitous or easily forthcoming. It must be noted that for all of the fanfare and fuss about Galileo as the father of modern science, his individual contributions to science were not all that many. He did invent a new telescope; he did "discover" the times-squared law. Much of what he did was issue promises for the future.

The promise of the unification of physics, a unification he did not himself achieve; the promise of a confirmed theory of the tides, a confirmation he himself failed to give. The historical and philosophical quest to discover and to understand his methodology must be based, then, not on the overwhelmingly large corpus of new discoveries made by Galileo (these were few), but rather on the confidence we want to bestow on what we take to be his having done science in the "right" way. The elegant and detailed historical investigations of Wisan and Shea and others; the careful and conservative philosophical analyses of McMullin and Wallace and others all seem to point to this way. Even Feyerabend, who thinks that Galileo was wrong on almost every count, nevertheless commits himself to the view that Galileo was one of the greatest propagandists on behalf of science ever to come forward, and he encourages us to imitate his ways of proceeding.[4]

Similar kinds of attention to the revelations of scientific methodology cannot fail to teach us humility. *For even if every putatively positive result of science is in fact negative – false – science is at least successful in that it knows the difference, and can show it to all who will listen attentively, learn actively, and observe with caution and with controlled vision.* From this perspective, methodology is the sought-after cipher: the pieces of the puzzle do not fit together because of a common, shared content, but because of an overlap in techniques, ways of going about doing the business of science. It is not *what* science says that counts; it is rather, that what science says can be known to be either true or false – that counts. Thus it is that an enlarged perspective on methodology reveals the often hidden epistemological well-springs of scientific work. Science pretends to tell us the truth about the world, pretends to vindicate Whewell's claim that "Man is the Interpreter of Nature, Science, the right interpretation."[5]

The epistemological questions to which we are led by a study of methodology are essentially questions of *appraisal*, questions having to do with why it is that, for the time being, we will accept the results of a given piece of scientific work. After all, there are plenty of reasons why we might expect a given result to turn out false; indeed, if we believe in induction as applied to the history of scientific discoveries, there is a high probability that it will turn out to be false. I need not in this place rehearse the view put forward by many that it is this very epistemological fragility of scientific results that argues for the health of science as a truth-telling institution. Wesley Salmon has recently argued that the historical success of certain kinds of hypotheses put forward by earlier scientists enters crucially into the probability of the success of a newly introduced hypothesis, by giving us a fix on the prior probability of the new hypothesis based on the past success of other hypotheses of the same kind – a success that yields plausibility to the positive outcome of the new hypothesis.[6] There is an ambiguity in Salmon's claim, a harmless and revealing ambiguity. "Hypotheses of the same kind" can mean many different things. Let me sort out a few of the possibilities. It can mean (1) "hypotheses with the same ontological commitments", or, (2) "hypotheses with the same *a posteriori* probability" or (3) "hypotheses introduced in the context of a methodology known to permit truth-value determinations with empirical integrity". Any of these hypotheses would satisfy Salmon's condition that the prior probability of a hypothesis is a relative frequency, the relative frequency of the success of hypotheses of the same kind. I will not detail the claim here, but it does seem to me that Salmon himself writes of hypotheses in all three of the above-mentioned senses.

What matters most is how we construe "success" in past hypothesizing and theorizing. Thus, although it is useful and perhaps illuminating to point out the three senses of "sameness", it is more to the point to see that the three senses of sameness are all of them required in order to begin to explicate what we mean by the success of a scientific hypothesis. Ontological commitments make a difference. It is important to note that we have more confidence in hypotheses that explain the motion of objects in this room in terms of commitment to physical objects, measurable distances and times, and calculable velocities and rates of acceleration, than we do in explanations committed to (as in the case of purported poltergeist activity) spiritual beings, nonmeasurable astral distances and times, noncalculable aspatial influences, and the like. Similarly, since the *a posteriori* probability of a hypothesis is a relative frequency, it seems quite legitimate to insist that acceptable hypotheses in science have at least as high a relative frequency as do any of those accepted in the past. Nevertheless, I want to emphasize that success in terms of *a priori* ontological plausibility, and success in terms of ascertained *a posteriori* probability both require methodologies and technologies that "work": in this case, work in the sense of providing epistemologically viable truth-value determinations. This means that the prior probability of a hypothesis is given by the relative frequency of the success of *the ways in which it is investigated.* The success of science is the success of scientific methodologies in solving our problems, no matter what end is served by any particular solution. If we suppose that, whatever else might have been the aboriginal intent of science, science means to tell us the truth about the world, to explain the world, to give us trustworthy ways of moving about in the world, then it is to the success of science as a way of solving our problems, *and* as an instrumentality in the orientation of our lives, to which we must turn attention. In this very specific way, epistemology also yields to methodology, and questions of truth turn into questions of success in practice, including, quite basically, among others, moral questions.

These provisional comments on a model of science as an institutionalized form of inquiry are meant to reveal to both historians and philosophers of science that their special interests are all of them fully justified by the character of institutionalized science, and that the fraternal rules of work in the two disciplines fully justify judgements of success or failure made entirely independently of any questions of satisfying aims larger than those circumscribed by the limits of each discipline. But beyond the special interests of those working within the two disciplines, lies an enriched context of the methodological and epistemological dimensions of science that are free of institutional control, and hence require us to take quite seriously questions of systematic intent: questions of aims and purposes. So regarded, science as a methodology becomes indistinguishable from its pursuit of this or that goal, and, paradoxically, the instrumentalities of science, its technologies, will again be seen to play a central role. These deinstitutionalized features of science, in other words, are only available to be seen when science as one kind of institution is seen as being in competition with other institutions. Perhaps it is no longer socially realistic to ask whether scientific work in institutionalized form should continue to exist, but it is certainly both reasonable and socially required that we ask how much science there should be, and beyond what point we should continue to support the institution.

In the pursuit of answers to such questions, the internal demands of institutionalized fraternal rules of procedure will not be satisfactory except to scientists and others who seek satisfaction of curiosity. For what is now involved is the examination of the credentials of science as a legal individual operating in a social contest against other institutions as legal individuals. This examination will have to lay stress upon the intimate connection between institutionalized science and technology, a connection so deep as to amount in some cases to an interdependency, if not a complete identification. The examination will also reveal a new social role, that of *scientific politician.* Science will need to be defended, but only members of the institution are fit to undertake the defense – only those who have mastered application of the fraternal internal rules will be able to defend the faith now that the institution is under social scrutiny.

What I am suggesting does most emphatically *not* invite replacement of philosophy of science and history of science by sociology of science. Those who apply my model will need to learn the social facts about science, but no replacement of one legitimate field of study by another is either required or envisaged. What I have been arguing for is a twofold thesis of the internal institutionalized priority of seeking satisfaction of curiosity in science and other disciplines on the one hand, and the continuity of methodology and epistemology, as defining properties of the functional activity of science, *viewed outside* the institution, on the other. Historians and philosophers, although they may not share much *in detail* in their work inside their institutions, working under accepted but different constraints, nevertheless share a great deal when their own activities, as well as those of the scientists, are seen as operating *outside* the given disciplines, and as competing with other forms of social activity, in any way at all.

Appraisals of the worth of a scientific product (a solved problem or a discovered "truth") can thus be undertaken in two fundamental ways. Appraisals of the internal success of a scientific result largely have reference to the fraternal rules of operation typical of the discipline. Although the rules of procedure are mainly conservative, a certain freedom of movement – realization of the urge to introduce novelty – is usually permitted. But what counts is that the result "solves a problem" generated by one of the prevailing research programmes in operation within the discipline.[7] Appraisals of the external worth of the scientific product always reintroduce expected utilities of the research, thus bringing into play various moral and other social cautions. At this level of appraisal, the technological dependency of science often carries a good deal of the burden of judgement.[8] Finally, the two kinds of appraisals direct attention to the two kinds of scientific success: *success within the discipline* (as measured against the demands of the prevailing fraternal methodologies, preferred ontological commitments, and the like); and *success of the discipline* (as determined by the effectiveness of the institutionalized science in providing solutions to other-than-scientific problems found to be troublesome in the social/political network of institutions operating at a given time).

The short form of my story is this. Both the historians of science/philosophy and the philosophers of science/philosophy have much to offer in any attempt to realize the benefits of my model. Historians can provide us with illuminating accounts of the "data" of appraisals; philosophers can provide us with analyses of the structure of appraisal, the "logics" of method and evaluation. Indeed, there is no reason why – in the best cases – the two tasks cannot be undertaken jointly, often even by the same researcher. But if the historian cannot, among other sciences, study philosophy, there is reason to doubt the reliability of his reading of the history of science/philosophy. Similarly, if the philosopher cannot do conceptual justice to the roots of his enterprise, there is reason to suspect that what appears arid and overly antiseptic in his work is just that – arid and antiseptic. All the rest is, for the most part, simply professional jealousy.

The University of Western Ontario

NOTES

* This is a much revised version of the paper communicated to the Pisa conference. I have tried to develop further some of the points discussed at the conference. I also hope that the paper has benefitted from discussions with Jürgen Mittelstrass during a visit to Konstanz following the Pisa meeting. Professor Mittelstrass had just returned from Aarhus, Denmark, where he read a paper entitled 'Towards a normative conception of the growth of knowledge' at a meeting sharing common interests with the Pisa conference. It was quite gratifying for me to observe that his study and mine arrive at roughly very similar conclusions, even though our work initially begins from somewhat different orientations.

[1] The reference to Aristotle is a mere historical courtesy. I am aware that the concepts of *experience* I am letting stand without proper explication are full of difficulties that cannot be overcome by simplifications of the sort I introduce here. Concepts of experience were the focal point of discussions in the *first* International Conference on History and Philosophy of Science, with papers by Jürgen Mittelstrass, 'Two concepts of experience: Methodological foundations of Aristotelean and Galilean physics', and by me, 'Experience and experiment as regulative principles in methodology'. Mittelstrass' paper has appeared in German as 'Metaphysik der Natur in der Methodologie der Naturwissenschaften', in K. Hübner & A. Menne (eds.), *Natur und Geschichte, X Deutscher Kongress für Philosophie, Kiel, 8–12 Oktober* 1972 (Hamburg 1973). I think we have here a very good example showing why it is better to have proceedings of earlier conferences appear *before* those of later conferences. It helps to keep both the chronology and the scholarship tidy.

[2] It will be obvious in what follows that I owe many intellectual debts. No one is responsible for my mistakes but me, but I do want to note that my thinking on these topics owes a great deal to what I have learned from Nicholas Rescher, in this connection especially his *Scientific Explanation* (New York & London 1970), and his *Scientific Progress, a Philosophical Essay on the Economics of Research in Natural Science* (Pittsburgh & Blackwell's 1978); from Wesley Salmon, 'Bayes's theorem and the history of science', *Historical and Philosophical Perspectives of Science, Minnesota Studies in the Philosophy of Science*, Vol. 5, ed. R. G. Stuewer (Minneapolis 1970); and from Larry Laudan, *Progress and its Problems, Toward a Theory of Scientific Growth* (Berkeley, Los Angeles & London 1977). Salmon offers an important way of conceptualizing scientific success; Rescher notes the absolutely crucial distinction between disciplinary science and knowledge-producing efforts not subject to institutionalization; and Laudan provides the most compelling reading of the *disciplinary success* of science so far available. No one of the three philosophers, however, would necessarily want to put their efforts at the service of my model, and in this respect Laudan will perhaps not be satisfied with my intertwining of various senses of 'success' in what follows.

[3] Larry Laudan, 'The sources of modern methodology', R.E. Butts & J. Hintikka (eds.), *Historical and Philosophical Dimensions of Logic, Methodology and Philosophy of Science, Part Four of the Proceedings of the 5th International Congress of Logic, Methodology and Philosophy of Science, London, Ontario, Canada* 1975 (Dordrecht 1977). The University of Western Ontario Series in Philosophy of Science, Vol. 12.

[4] For recent work on Galileo, see, for example, the relevant papers in R.E. Butts & J. C. Pitt, (eds.), *New Perspectives on Galileo, Papers Deriving from and Related to a*

Workshop on Galileo held at Virginia Polytechnic Institute and State University, 1975 (Dordrecht 1978). The University of Western Ontario Series in Philosophy of Science, Vol. 14; for P. K. Feyerabend see his *Against Method, Outline of an Anarchistic Theory of Knowledge* (London 1975).

[5] As I shall be trying to argue in what follows, the standard epistemic predicates 'true' and 'false' lose their integrity when epistemology is thought of as intimately related to methodology. If, with Laudan and others, we take the decisive unit of scientific success to be the problem, then perhaps 'solved' and 'not-solved' will have to replace the standard epistemic predicates. It is curious to note that the epistemic predicates appear to take precedence over other utilities in relatively methodology-free philosophies of science. This seems especially true of the philosophy of science traditionally associated with Hume, and of recent positivism. I have argued that it is appropriate thus to disassociate science from its methodologies for purposes of studying the logical features of scientific systems. But this disciplinary abstraction *is not science*, if what we mean by 'science' has reference to an ongoing historical enterprise. I would suggest that the larger the methodological perspective becomes, the smaller is the true/false content of science seen to be. In any case, it seems to me that the cardinal emphasis in science is not upon true/false decisions *per se*, but simply upon the fact that scientific outcomes provide decisions of some sort. And it is the methodologies, not the abstract (reconstructed) epistemologies, that tell us what will count as a decision.

[6] See the essay by Salmon in note 2 above. I have employed Salmon's thesis in another context in my 'Consilience of inductions and the problem of conceptual change in science', in R. G. Colodny (ed.), *Logic, Laws & Life, Some Philosophical Complications* (Pittsburgh 1977); University of Pittsburgh Series in the Philosophy of Science, Vol. 6.

[7] This is the prevailing theme of Laudan's *Progress and its Problems.*

[8] For a detailed and enlightened discussion of the idea of the technological dependency of science see Rescher's *Scientific Progress*, 132–192.

J. D. NORTH

ON MAKING HISTORY

When these meetings were being planned it was my impression that the aim was to bring together members of the intersection class, so to speak, of the history and philosophy of science. We should, as I thought, be a priesthood advocating – despite our own shortcomings—the marriage of history and philosophy of science, and deploring divorce, unlawful cohabitation, and associations of a kind it used to be polite not to mention. Things are not turning out quite as I expected. Some of our number clearly want to defend the single life, while others put their faith in shotgun weddings. Since in much of what follows it will seem that I am playing the part of spectre at the feast, let me say at once that I do believe in the logical possibility of marriage, as long as it is of the right sort. This morning I shall speak only of the tensions in marriage, and not of the intrinsic worthiness of the partners.

We are all of us concerned with different types of explanation—with their feasibility, their legitimacy, and even with their relative worths. Lest you doubt this last point, let me remind you of W. Stegmüller's characterization of a type of understanding which involves knowing how people react to a situation because one is a person oneself. It is understanding based on "judgements concerning spontaneous human reactions" and is "a typical case where acquiring understanding remains superior to any attempt at historical explanation".[1] Not just *different*, but *superior.* I can sympathise with him at this point. As human beings we are, as he says, competent to evaluate certain reactions as "typically human" without having to resort to generalizations with a hypothetical character. Even historians are human—but they are human enough to know that human beings change from century to century no less than from hour to hour, and that historical explanations can be given of these changes. I shall touch on aspects of this problem again, but I shall do so only in the context of potential liaisons between Moulines' style of philosophy and writings on the history of thermodynamics.

As you know, Dr Moulines writes in the Sneed-Stegmüller tradi-

J. Hintikka, D. Gruender, and E. Agazzi (eds.), Pisa Conference Proceedings, Vol. II, 271–282.

tion. He looks beyond the statements of the sciences, as such, and beyond the specific theories of the sciences, to the general structures which those theories might share. His ambition is to remove vagueness from discussions of structure, and to do so by axiomatization. He uses the method of defining set-theoretical predicates (which has formerly been used to clarify the structure of individual theories) to clarify meta-theoretical structures. The metaphorical nature of such words as "structure" and "theory-frame" (perhaps better "family") as they are used in general discussion is easily forgotten in a sea of set theory, as is the fact that similar discussions (less strict but more realistic?) have elsewhere had a motivation which was primarily historical. Families, in the historian's world, include infants and old men alike. To forget this when we speak of families of theories is to submerge a vital historical distinction. The discussion of historical change is not new to historians, however loose the language of that discussion. Despite the fact that one of the prime aims of the "rational reconstructions" of recent years has been to show what it is to be a paradigm (in something like one of the Kuhnian senses), these reconstructions have a rather unhistorical look about them, which is not much altered by the inclusion of the parameter t here and there. They usually seem to require, for instance, that the theories which share a structure have shaken down, as it were, so that all who work with them (or a discerning outsider?) are agreed on what it is which allows them to be grouped together (so as to qualify, for instance, as "theories of equilibrium thermodynamics"). This is calculated to make the historian of these theories uneasy. He is likely to say, in effect:

"If you are going to wait for the dust to settle, and look only to those periods of history at which there is a consensus about what constitutes (for instance) thermodynamics, then you will automatically rule out much that the historian finds interesting. You will read history backwards. You will look only for those historical phenomena which fall into your logical scheme, and ignore – even disparage – all else".

We are thinking here, of course, of Moulines' level 3, rather than of such axiomatizations as Joseph Sneed's reconstruction of classical particle mechanics. The historian might also be uneasy with the latter approach, despite its lower level of generality, on the grounds that it picks out elements from quite different historical epochs. (One can hardly "sign" each of the concepts in the axiomatization with a time parameter giving the date of its introduction!) At both levels, the analogy with the empirical sciences is a seductive one, even though most of us have more confidence in the uniformity of the world than in the uniformity of historians, that is, of those very human human beings who provide us (the division of labor being what it is) with our historical data. And at both levels there is the old problem of the *uniqueness* of the historical past, which is what the historian will say he is trying to understand. It is not the same thing, he will say, to understand the evolution of Clausius' thought as to impose upon the overall pattern of his thought a set of axioms (or several sets) which it happens to satisfy. "Don't worry", the axiomatizer might reply, "I am trying, not to catch every aspect of Clausius' thought, but only its general structure, and perhaps also the laws governing its growth. I shall leave to you the task of filling in further historical details". By "historical details" it must be admitted that the philosopher often means whatever he really believes to be trivial—or at best, whatever is relevant to types of explanation other than that he believes he can himself provide. The phrase might refer, for instance, to the precise wording of a quotation, to the color of the ink in which it was first committed to paper, to a scientist's religious life, or to other aspects of his work which are no longer regarded as scientifically reputable. It might even be considered that the logical order of presentation of a theory when it was first announced, or the empirical basis it was then supposed to have, is irrelevant to the deeper historico-philosophical meaning of that theory. Even among philosophers of science there are many shades of autocracy.

I hardly need say that these are topics in the perennial debate between historians and philosophers of history. There are many ways of characterizing the two professional groups, but the philosopher tends to caricature the historian as a man entangled in the minutiae of unarticulated (and hence boring) facts, while the historian sees the structuralist philosopher of history as a naïve and excessively optimistic soul, who has lost his way to the geometry classroom, and whose limited historical experience prevents him from appreciating the infinite complexity of the history of human thought and actions. Of all professional historians, the historian of mathematics and science is likely to be the least hostile to attempts to impose formal structures on history, but even he will be inclined to despair. He will, like all historians, be much concerned with historical change, and he will emphasize (often if only to display his own refined powers of perception) the subtlety and frequency of transitions of thought. Even if he were to grant that a man's thought could be represented axiomatically (with due rejection of historical detail), he might well argue that the effort is not worthwhile—that Einstein's shifting thought, shall we say, might have needed ten axiom systems before breakfast, and that the redundancy they would involve us in, not to mention the sheer pointless effort, cannot possibly be justified.

The historian's deep concern with historical change makes him almost paranoid about anachronism, and the first specific charge he will bring against the structuralist philosopher of science is that of reading history not only backwards, but almost randomly, and of appropriating words so as to impose on them meanings they did not always have. A case in point would be Rankine's "thermodynamic function", later seen by Rankine himself to correspond to Clausius' entropy function. An axiomatization would immediately display (in the absence of some sort of "footnoting") the sameness of the two concepts—even though at the time of its introduction (1854) the role of Rankine's function was very dimly appreciated. No true historian of thermodynamics could be content were this fact lost to view, and most would feel that it is a fact more suitably recorded in prose—or even in verse—than in the profundities of set theory.

The danger of appropriating concepts in an unhistorical way is present at both level 3 and level 2, in Moulines' sense.[3] According to him, for example, all theories qualifying as "thermodynamic" are assumed to satisfy conditions (a) to (g) of his section IV.2. The inspiration for this list of conditions comes from simple equilibrium thermodynamics, for "it is plausible to think that simple equilibrium thermodynamics should be taken as a paradigm for the other, more complex theories in the sense that the essential structure will be the same"[4]. Historically speaking, the word "thermodynamic" seems to have been first used by Thomson in 1840 merely to imply a relation between heat and work. This was before he had given up Carnot's approach and adopted the "dynamical" theory of heat (i.e., when he was using the concept of heat content and not the kinetic theory), and it is ironical that this was all before Thomson was in possession of the concept of entropy. Now Thomson's stance vis-à-vis the kinetic theory is, on Moulines' analysis, of no importance, for Thomson's is a theory which I gather Moulines would place outside thermodynamics (see his (d), which rules that the only thermodynamic concepts are energy, entropy, and the monadic property "is an equilibrium state"). The concept of entropy was not, as I have said, in Thomson's early work; but what if we find that it can be *superimposed* on Thomson's theory (by definitions drawn up in terms of functions which are explicit in his early work) without any crucial emendation of that theory? Is the early Thomson theory to be classified with later "thermodynamic" theories? And if so, has there been any gain in historical insight as a result of this classification? I have a vision of an aeronautics engineer making a rational reconstruction of the Wright brothers' biplane, and finding when he tries it out that it flies at the speed of sound.

I realize that not all of us will wish to give the same answer to my question about historical insight. Those who look for formal qualities in history—that is to say, by and large, those who do not practise history—will argue for formalization, while those who view history as a sequence of unique and unrepeatable events, and who find the richness of the texture of history more important than hidden patterns in it, or than patterns into which (as they will say) it can be forced, will tend to disagree strongly. The split in temperament is as old as history itself; and even in the formalised disciplines there have always been marked differences of opinion as to where the process of generalization should stop. How can we reconcile such fundamental differences of opinion? Not by preaching *in abstracto*, but by being scrupulously severe about matching up our "rational reconstructions" of history to the best historical data. To do less is only one stage removed from the rationalist's *a priori* construction of the physical universe without reference to empirical findings. And this parallel at once underscores the problem of the involvement of theory in data, historical data no less than the data of science. Does this not mean that those responsible for the axiomatization of history must—if they are not to gather their own data—at least become sufficiently familiar with the proclivities of "ordinary" historians to make allowance for the "personal equation", so to say? It seems to me obvious that no historian will even take them seriously unless they do so.

Another problem analogous to one faced by the rationalist metaphysician is that of excessive generality—or, as a critic would say, of vagueness. Dare I draw a parallel between Kant's Principles of the Understanding and Moulines' axioms? As you know, the circumspect Kant spoke of the understanding as both a "power of formulating rules through comparison of appearances" (as in the natural sciences) and as a "lawgiver of nature", conferring upon appearances their conformity to law, and so "making appearances possible" (*KdrV*, A 126–7). Reason was then said to be the judge who compels the witnesses to answer questions which he himself has formulated. For "reason" read "rational reconstructors" and for "witness' read "historian of science". I see no problem here. For Kant, reason has a plan which *must* be followed, not by Nature but by knowledge of Nature. I take it that the axiomatizations which the future might bring at "level 3" will not be deemed to have the same necessity as far as historians of science are concerned. Does Kant not serve as the awful warning that is needed here? His principles had almost no influence on those whose task it has been to extract the (lower order) laws of nature, and in fact there are many who would maintain that Kant's views on space and time have acutally inhibited creativity in corresponding branches of mathematics and physics. In other words, when his philosophy was specific enough to interest the scientist it stood or fell on its scientific merits, and when it was general enough to be safe it was *uninteresting* to the scientist. I must apologise if all this sounds too much like a morality tale, but the analogy with Mouline's axioms seems to me a close one. I simply doubt whether his axioms are ever likely to influence or interest most historians. As he himself admits, "On the whole not many interesting conclusions can be drawn from the axioms of an operational basis because of their generality".[5] If a philosopher can say this, no historian is likely to disagree.

Another objection likely to be raised by the historian concerns the *insensitivity* of the axiomatizations (at all levels), which might even escape the worst forms of anachronism and yet give consistency to a theory which originally lacked it. It would be a bold philosopher who attempted an *irrational* reconstruction of the history of scientific development, but this might be the honest thing to do. Very occasionally the scientific debate itself forces inconsistencies on our attention. (Take, for instance, Poincaré's theorem of quasi-periodicity with which Zermelo hoped to demolish statistical mechanics, whereas Einstein used a very similar procedure to draw a wholly different conclusion. Whether or not there was any strict logical conflict between the two parties to the dispute, I can see no axiomatization on the horizon capable of doing justice to the situation.) For the most part, historians and philosophers of modern science are the embodiment of kindness and charity towards scientists, and prefer to give them the benefit of the doubt as regards this question of consistency (or perhaps I should say "avoidance of inconsistency"). Perhaps this is only a sign that inconsistency is consistent neither with hero-worship on the one hand nor with a satisfying axiomatization on the other.

Another quality most of us will demand in formal representations of the historical process is conciseness. It is hard to contemplate those axiomatizations of the future which far exceed in length the raw data which they are held to encompass; which lead us to conclusions (with exactness and confidence, it is true!) that had been proposed long before, albeit in the form of untidy sentences; and which indeed were framed so as to lead to those conclusions. "But", you can hear the future historian saying, "I knew all along that psychological factors play an important part in the genesis of scientific theories". "No", some future structuralist might reply, "you only *suspected* that the psi-factor could be introduced into meta-history. I can now show quite precisely how your suspicions were, broadly speaking, along the right lines".

This is clearly not enough. The structuralist's scheme must do more than give a formal guarantee of the coherence of historiographical principles already in vogue. Like the empirical sciences, it must reveal the unexpected if it is to be taken seriously. I am not qualified to comment on the present state of "structuralist" philosophy of science, but I confess to having been disappointed on this score. In the explication of so-called "normal" science, for example, we have an explanation of why it is reasonable to treat theories as safe from refutation at the hands of contrary experience. But was this point not being made by Poincaré, and Eddington, and others, more than half a century ago? Again, according to Stegmüller, normal scientific progress is definable in terms of "cumulative development of Kuhn-theories", while (as he goes on to say) "in progressive scientific revolutions 'the displaced theory can be partially and approximately imbedded into the supplanting theory'".[6] And yet discussions of "approximative imbedding in the supplanting theory" are not rare in historical and physical writing. Yet again, when we recall recent discussions of the question as to whether the decision to reject a theory is simultaneously a decision to adopt a *new* theory, we tend not to be astonished, but for a different reason. No longer is it the case that we have heard the answer before—now we can see that there are many possible answers. For, however tightly we define "adhering to a theory", it will never be positively illogical for individuals (or groups of individuals) simply to walk away from the entire subject. There are numerous historical examples of the "emigration syndrome", in fact. Of course, statistically speaking it might prove to be insignificant, and an analysis in terms of community inertia might be much more useful. I do not wish to be sidetracked for the time being into considering the adequacy of current "structuralist" concepts, but merely to draw attention to the present unreadiness of the structuralist analyses for that continuing search for refutation which is usually taken to be a mark of the critical spirit.

The suspicious historian is likely to be delighted at such a state of affairs. We can imagine him drawing an analogy between the astronomer, unable to influence the properties of the distant celestial objects he studies, and the structuralist, unable to affect the unique flow of history. But the analogy with the empirical sciences (an analogy I have used perhaps too often) is a dangerous one when we are concerned with a subject which is largely sociological, and therefore, as it might be crudely expressed, "ontologically poor". There *is* an analogy with the cosmology of the nineteenth century here, as it happens—namely with a cosmology in which no one was quite sure how the nebulae were to be classified. There was no similar disagreement about heat baths, steam boilers, and sawdust. As befits a set of concepts which function at philosophical and sociological levels, "progress", "community", "paradigm", "scientific revolution", even "theory" and "person", are not readily interpreted. Disagreement over the pre-formalised use of such basic concepts, combined with the coarseness of the mesh of the axiomatizations incorporating them, might put the possibility of "historical test" far in the future, but does not rule it out of court. (I suppose some philosophers will say that as soon as structuralism finds itself empirically tested, it will cease to be philosophy. *Per contra*, there will be those historians who use its key phrases because they are useful in structuring historical material, and possibly others who do so because those key phrases have a certain cachet in the academic world.) If I seem skeptical of the possibility of historical test, it is not as a matter of principle. It is rather that it seems to me that much more groundwork remains to be done in liaison with historians of science.

This must all sound both obvious and pretentious. I know better than anyone how pretentious it is, for I have neither a good grasp of the work that has been done already nor a programme for the future. I must be content with a few final disjointed ideas. It seems to me that it is unfortunate that attempts are being made to "rationally reconstruct" two things at once. The new movement seems to have begun with attempts to tighten up physics (as Euclid, in the old story, tightened up the rules of the carpenters), or to put physics on a sound operational footing. For these purposes there is nothing illogical about defining heat (for example) in terms of entropy, which was historically second on the scene. The temptation was then to move to Moulines' level 3, where the entities are whole theories. The step was original and intrinsically interesting. However, complete theories are *not* the fundamental entities of history of science, as one is tempted to believe when hypnotized by the delights of axiomatizing restricted physical theories. Believing that theories are basic leads to a disregard for the early and tentative phases of science. By laying down criteria for a theory to qualify as a theory of thermodynamics, we *force* upon the historian an abrupt transition from the period before there were any "truly thermodynamic" ideas to the period when there were theories which can plausibly be said to have satisfied the criteria. It seems that scientific revolutions and periods of normal science are merely reflections of our desire to pack history into a few neat boxes—and to pretend that what cannot be boxed does not really matter.

The attempt to extend the axiomatization of physical theories to an analysis of history, seen only as a history of theories, has other unfortunate consequences. To theories is assigned the job of uniting individual scientists into the same tradition. (There may be other ways of grouping scientists together, allowing more scope for heresy. This fact is beside the point.) I gather that some writers have been led from this position to one in which the concept of a scientific community is taken as basic. In neither case is the historian of ideas going to be content. Not all history is sociology. Many historians of ideas—I would say almost all—prefer to start with the individual; with the individual responsible for innovation; with the individual whose beliefs are more easily grasped than those of the group; with the individual who holds a pen; with the individual whose ideas may lie fallow for generations, or who may be a century behind his time, and yet still historically significant; with the individual whose personal qualities were such that his beliefs were almost ritually acknowledged as valid. Even at the level of individuals, any attempt to formulate a calculus of belief is likely to run into difficulties, unless it is left in terms so general as to be without useful historical application. Different individuals, especially from different historical periods, are prone to offer different sorts of reason for holding to their beliefs. Such difficulties are not new to sociology, where the thoughts of individuals are customarily smoothed out by approximative methods. Few historians will accept the loss of the fine-structure of history which such methods entail. One might compare the recalcitrant historian with a painter who objects to Euclid on the grounds of his failure to take red lines and blue lines, irregular lines and blurred lines, into account. I cannot see that Euclid would gain anything by affirming and reaffirming that painting is nothing but geometry, or even that it is "essentially" geometry. He might teach the painter something; but in the end, the two must go their separate ways.

My analogy with Euclid and the painter is of course more appropriate to the bad old days when philosophers were chiefly concerned with a "statics" of scientific theories than to our own enlightened times, when the kinetics of theories adds an extra philosophical dimension. The advance is a significant one, but it is not likely to affect the historian very directly. The logic of science, whether a statics or a kinetics, can in turn survive without the historian's help, as long as its practitioners confine their attention to the realm of the possible. But if the two subjects are to interact, it seems to me that the greater influence by far must be from history to philosophy of science. It is easier to act the part of Cassandra than to give precise details of the influences we are to expect, but I can at least give an example of what I have in mind. To take Rankine again: here we have the historical phenomenon of two largely independent traditions—his and the Clausius-Thomson tradition—being developed independently. Broadly speaking, we can say that Clausius and Thomson derived Carnot's law (of the efficiency of heat engines) from the principle that heat cannot, by itself, pass from a cold to a hot body. Rankine, on the other hand, who was for long ignorant of the details of Thomson's proof, and who disliked that given by Clausius, derived the law instead from his hypothesis of molecular vortices (1851, etc.).[7] It was almost twenty years before Rankine abandoned his vortex hypothesis (essentially one which required that absolute temperature is proportional to the square of the vortical velocity of atomic atmospheres around their nuclei) and acknowledged the more conventional kinetic theory of heat. No "sociologist of knowledge" can define away the intellectual and practical influence of Rankine, who must somehow be fitted into the categories of any *historically* acceptable analysis of the kinetics of theories, simultaneously with Clausius and Thomson. Faced with the problem, the structuralist philosopher of science, new style, might incorporate into his schemes a branching (and perhaps a rejoining of branches) from revolutionary nodes, with consequent simultaneous traditions of "normal" (or "extraordinary") science. No doubt there will then be repercussions on the discussion of the concept of progress. No doubt there will be repercussions on the very notion of scientific theory, and scientific community. The idea of divergent traditions seems to fit perfectly well, in fact, with Moulines' idea that a theory of thermodynamics is characterised by (among other things) a parameter *not* presupposing thermodynamics. He would, therefore, presumably distinguish between Rankine and Thomson on the grounds of their different choice of non-thermodynamic parameter. Whatever the possibilities for combining these ideas, they seem to me far stronger than the possibility of a significant feedback of anything more than a few simple philosophical categories and concepts into the historian's world. I hope that history will prove me wrong!

Any such feedback, from the philosopher's total system, is likely to be delayed and indirect. Even so, it is potentially extremely valuable. I am not very sanguine about the axiomatization of painting, but painter and critic alike will not be the poorer for Euclid's structures. Much the same goes for history and the categories of the structuralist philosopher of science. It would be unrealistic, however, to expect too much. The artist is not likely to have much use for the nine-point theorem; and if some zealous philosopher should one day axiomatize a Richter scale of scientific upheaval, he would not only take away most of the fun of courses on Scientific Revolution (attended largely, it seems, by students attracted by the simplest of political analogies), he would almost certainly be ignored by historians. This would not entirely be because historians are simpleminded. It would in part be a consequence of the fact—often overlooked, it seems, in the recent debate over the nature of scientific revolutions—that the historian's categories are often taken from systems of thought within the very societies he is studying. Revolution, renaissance, progressive—the historian and the philosopher might be dissatisfied with, criticize, and revise ways in which these concepts were, at the very time of their invention, applied to contemporaneous events; but it would be wrong to ignore those past applications, which are themselves an ingredient of history. Even in history there is interaction between the observer and the observed.

Rijksuniversiteit Groningen

NOTES

[1] 'Accidental (non-substantial) theory change and theory dislodgement... [etc.]', *Erkenntnis* **10** (1976), 165.

[2] 'An example of a theory-frame: equilibrium thermodynamics'. This volume.

[3] *Ibid.*, p. 217.

[4] *Ibid.*, p. 219.

[5] *Ibid.*, p. 233.

[6] I. Niiniluoto, 'The growth of theories: comments on the structuralist approach'. This volume.

[7] For a comprehensive treatment of Rankine's thermodynamical ideas see the D.Phil. thesis by K. R. Hutchison (Univ. of Oxford, 1975).

L. A. MARKOVA

A COMMENT ON J. D. NORTH, 'ON MAKING HISTORY'

J. D. North speaks of historians' reaction to attempts made by T. Kuhn jointly with Sneed and Stegmüller to formalize the interpretations of the revolutions in science. In some instances North successfully formulates the obstacles encountered in any formalization of history. His view applies, moreover, not only to formalization, but to any logical and rational comprehension of the past. A logician or philosopher, North duly observes, primarily proceeds from contemporary interpretations of one concept or another. Thus his approach to the past is shaped by his willingness to find something in the past resembling the present. The historian's most important goal is to treat past events in their inherent meaning, irrespective of their significance today: that is, to view them in their uniqueness as being characterized by some individual features which they do not share with other events.

J. D. North mentions another weakness of any formalization: it covers something already available, some concept of the historiography of science approved and recognized by historians. Thus, formalization excludes unexpected findings and discoveries. The historical knowledge already available is reformulated with accompanying loss in historical specificities; it is not clear what this formalization is needed for. The formal logician's research generally leads him to nothing but common knowledge.

Yet, North's approach in criticizing attempts at formalization is, in essence, that of a classical historian, which is ever subject to refutation in any effort at philosophical or logical comprehension of the past, opposing such comprehension to a historical description. North merely reproduces these positions, in no way emphasizing them. Undoubtedly useful is the very reminder of some specifically historical values which have retreated into the background today – if not further – pressed by the generally-felt desire to make history rational, philosophical and conceptual. Still, such a "mere reminder" in no way

J. Hintikka, D. Gruender, and E. Agazzi (eds.), Pisa Conference Proceedings, Vol. II, 283–285.

eliminates the problems that emerge from the critical obstacles in a historian's work with a "purely historical" approach. The historian, North states, is interested in the individuality of past events and, while analyzing a certain epoch he uses concepts in their meanings of those times. However, before the historian of science engages in studies in the uniqueness of some past event, he has to be prepared to select an event or events belonging to the history of science and not to, say, the history of philosophy or political institutions. To do this, he must have his own opinion as to what science is. Suppose he proceeds from the understanding of science accepted during the period under study. But can we say unambiguously how the concept of science was understood at a certain time? Whose testimony are we to account for: that of philosophers, scientists, politicians, or common folk? Further, could scientists or philosophers of a certain period agree on what science was? Then, whose opinion is to be taken as that of the epoch?

In answering these and similar questions, the historian, irrespective of his own wishes, is hardly in a position to avoid evaluation of past events from his contemporary standpoint. Generally speaking, is it necessary and possible at all for one to escape oneself? Can the uniqueness of the past be preserved by complete elimination of the researcher's "uniqueness"? Let us consider the limitation often imposed by historians to the effect that past events be evaluated with regard to their significance for their time rather than the present. Many of the scientific discoveries made "prematurely" did not affect the scientific research conducted in that epoch. But what about a historian who fails to mention, for example, Mendel's discovery among the fundamental discoveries of his time? Such a historian is sure to have made a mistake, but could he appreciate Mendel's discovery other than from the standpoint of later history? Thus, in quite a number of cases the very need (and the prerequisite of a historian's work) to judge the past from the viewpoint of the present seems to be quite justified purely historically.

The demand that a historian of science should study events from the point of view of their uniqueness is not as evident as it might seem at first glance. We have already mentioned the fact that the very choice of an event presupposes some general knowledge. Further, a unique event investigated and reconstructed in some way or other is to be associated either with the overall context of social life (and then it will be incorporated as a scientific event) – again one needs the notion of science – or it will be viewed in the context of scientific progress alongside other unique scientific events. In the latter case, the question inevitably arises as to what general aspects make it possible to treat them collectively as scientific events, and what is general, not unique, about them? It looks as if the historian cannot avoid generalizations in looking for the general features characteristic of a series of unique events, no matter what his devotion to uniqueness might be.

We see thus that North's approach provokes at least as many questions concerning the historian's professional work as that of his opponents: philosophers and logicians. One cannot but appreciate the value of North's report in his description of contrasting situations in research in history characterizing the standpoints taken by historians, or philosophers and logicians Yet, North is rather narrow in seeing obstacles to research in history in the logical approach alone; whereas, as we have attempted to prove, similar hindrances are inherent in the traditional approach accepted in history.

CARLOS-ULISES MOULINES

REPLY TO JOHN D. NORTH, 'ON MAKING HISTORY'

I guess there are many points in Prof. North's *Comment* I should try to deal with. However, for brevity's sake I shall concentrate on two of them only, *viz.* the question of history's uniqueness and the objection that the kind of logical reconstruction I am advocating considers scientific theories as finished products only – something which apparently is of no much interest to the historian of science. These two points seem to me to summarize best North's doubts about the possible relevance of the philosophy of science to the history of science. But before saying something concrete about them I would like to make a general comment on the issue of the interaction between philosophy of science and history of science.

I would not like to discuss this issue in terms of the mutual influence that different professional groups (in this case, historians and philosophers) can exert on each other. This might be of interest to the sociology of the academy, but it is of no concern to me now. I am, instead, interested in the possible relationships between two groups of conceptual tools for analysis of science that apparently belong to two academically distinct disciplines logico-philosophical reconstruction (or, if you prefer, systematic philosophy) on the one hand and historical analysis on the other.

No-one seems to doubt nowadays that historical analyses of particular facts about science might have some bearing on the philosophy of science. At least, I do not doubt it, nor does North seem to doubt it. The real issue at stake is whether the converse relation may also hold in any interesting sense. I wish now to argue in favor of this possibility.

I think that logical reconstructions of given portions of science can be relevant to the study of the diachronic aspects of science in more than one way. The different ways in which this might happen should be clearly distinguished before we try to reach sweeping evaluations of the relationship between the philosophy of science and the history

J. Hintikka, D. Gruender, and E. Agazzi (eds.), Pisa Conference Proceedings, Vol. II, 287–290.

of science. Let me mention just some ways in which the "converse relation" might hold.

First, philosophical (possibly, in addition, formal) analyses of science can be relevant to particular historical case studies where logical and methodological issues are clearly at stake. The formal tools and the controlled terminology of the philosophy of science can be of much help when we try to conceptualize and interpret some particular piece of scientific development in a cogent way. There are plenty of examples of work done along these lines. Let me mention just three contributions that have been discussed in this Conference: Hacking's work on the emergence of the probability concept, Suppes' reconstruction of Archimedes' axiomatization of statics, and Krüger's analysis of Boltzmann's alleged reduction of thermodynamics to statistical mechanics.

Secondly, there is the possibility of casting a wider look at the development of science while trying to detect some implicit general patterns fitting a philosophical metatheory we have conceived. I am thinking, here, of the historico-philosophical conceptions of authors like Kuhn, Toulmin, Lakatos, and Laudan — to mention just a few recent examples. What we have in these cases partly belongs to history, partly to philosophy; in any case, its essential component consists in devising a philosophical theory of science which is then used to interpret the history of science. One might object that the temptation to undertake this kind of philosophy of the history of science should be resisted, since it is too risky — it can lead us astray. But then we should avoid philosophy altogether, for interesting philosophies are always risky: they can always lead us astray.

If we admit the legitimacy of this philosophizing about the history of science, then we may also be more ready to accept a third sort of connection between philosophy and history. Since the general philosophical views of scientific development just mentioned are normally stated in rather vague and controversial terms, an effort towards formal clarification, leading to a better interpretation and distinction of the issues involved, might appear worthwhile. Many misunderstandings and obscurities can be dispelled in this way. (It seems clear to me that the use of precise concepts is more likely to contribute to the understanding of philosophical issues than the use of vague notions.) An example of this kind of work, namely a formal systematization of philosophical views on scientific development, is the structuralistic reconstruction of, say, Kuhn's or Lakatos' philosophical schemes. In my opinion, this is an indirect, but important, sense in which formal, systematic philosophy of science can be relevant to the diachronic analysis of science – but it is certainly not the only possible way.

Finally, we might try to devise general and exact metatheories of science with no apparent bearing on historical case studies, reconstructing only what North calls the "statics" of science. I think they can also play an important role in connection with the study of actual scientific development because they provide, so to say, the ideal patterns or logical schemes by means of which we can measure the distance between actual and ideal scientific products.

It is from this point of view that the problem of historical uniqueness can be best understood. In fact, it is my contention that the talk about historical uniqueness would make little sense unless we are equipped with such ideal patterns as provided by a general metatheory of science. It is only against the background of some ideal types of scientific structures that the uniqueness of actual concrete scientific products can be assessed and understood. I suspect that most people interested in *understanding* the actual development of science do have such ideal types in the back of their minds in a more or less implicit way. To make such ideal types explicit can be a task for the philosophy of science. The notion of a theory-frame as introduced in my paper, for example, could be a useful tool for that task,–at least in the case of thermodynamics. The thermodynamic theory-frame as a family of theories is certainly an ideal kind of entity which we can expect merely to approximate actual scientific theory-building to a certain degree. But, at the same time, it is this kind of entity that would provide a clear meaning to the statement that Rankine's or Thomson's original ideas have a dubious or fuzzy status with respect to (present day "finished") thermodynamics.

This leads us naturally to North's second main point: his objection that philosophy of science, and particularly the structuralistic approach, deals only with theories as finished products. I think this would be a real objection for someone who contends that only finished theories should be considered either by philosophers or by historians of science, but not to someone, like myself, who believes that finished theories *might also* be of interest at least to some philosophers and historians. As far as I can see, this interest can be due to two different sorts of reasons:

(1) because finished theories can be viewed as ideal types against which to compare the actual flux of science;

(2) because finished theories are not only ideal entities but also often appear as real entities in scientific textbooks: they too are *historical products* in a period of Kuhnian "normal science".

Let me now concentrate on reason (2). If we accept the existence of normal scientific theories as historical products, among others, and if we accept that the logico-philosophical reconstruction of conceptual products in general is a sensible undertaking, then we can also accept that the logical reconstruction of pieces of normal science as finished products not only makes sense but is also relevant to the history of science proper (since a *part* of the history of science is *also* the history of finished scientific products).

Let us call this study "the philosophy of scientific textbooks". Contrary to a very widespread opinion, I think that this kind of study is both legitimate and valuable for philosophers and historians alike, but it has certainly been neglected by the recent discussions of scientific change. For some reason or another, people seem to feel more enthusiastic about heroic science.[1]

To conclude, my goal was to contribute in some way to the philosophy of textbook science in the concrete case of thermodynamics, – both in order to reconstruct real products of a given scientific development, and to provide an ideal framework as a background to evaluate the more unique and dramatic aspects of science. In both these ways, the study of the thermodynamic theory-frame might be, in principle, an example of a logico-philosophical reconstruction having some bearing on historical interpretations of science.

NOTE

[1] I use the term 'textbook' here in a broad sense as an abbreviation for "finished scientific product commonly accepted by the scientific community". This includes not only introductory standard textbooks but also advanced, systematic works on a subject. For example, in the case of thermodynamics the term would include Callen's *Thermodynamics* and Tisza's *Generalized Thermodynamics*.

L. V. CHESNOVA

INFLUENCES OF SOME CONCEPTS OF BIOLOGY ON PROGRESS IN PHILOSOPHY

Cooperation between sciences in their development process is among the most important issues of research carried on in history of science. Another highly significant problem refers to the forms and consequences of their interaction.

The history of interacting methods of research deserves particular attention in an analysis of interrelations and mutual influence between biology and philosophy.

It is only natural that the experimenter and the philosopher understand the term 'methodology' in somewhat dissimilar ways. For the experimenter, a method means guidance in his experimentation activities. In due time, having been repeatedly verified by practice, a method may find still broader application, and gradually become a certain style of research.

The philosopher, however, is attracted by the theoretical implications of a method as a style of thinking. When he analyses this category, he starts with the role of the method in the cognitive process, then he isolates the general idea to be used in the formulation of the principles underlying research activities.

The present communication deals with this angle of mutual influences exercised by history of biology and philosophy. Certain relations between the basic concepts of philosophy, on the one hand, and evolution theory and the experimental method in biology, on the other, are quite evident. However, in the former case, the emergence of a fundamentally new concept of evolution in biology directly caused positive results in the promotion of progressive trends in philosophy, whereas in the latter case, the experimental method, paradoxical as it may seem, contributed to the revival of vitalism.

It is a well-known fact that A. Comte, the founder of a school of philosophy widespread in the West, stated that a 'metaphysical concept of the impossibility of reaching absolute knowledge' is to be

J. Hintikka, D. Gruender, and E. Agazzi (eds.), Pisa Conference Proceedings, Vol. II, 291–293.

the methodological foundation of this school. According to Comte, science is to restrict itself to collecting empirical data, describing facts and preparing their inventories. According to the positivistic doctrine, the utmost aim of science consists in formulating outwardly evident relations between phenomena. This highly specific interpretation of the category of law is an almost complete neglect of causal relations between phenomena; in other words, the utilitarianism characteristic of Compte's philosophy could not orient science towards truly profound research.

Ch. Darwin's theory, which smashed the fundamental doctrines of positivism, in fact, became the antipode of this method applied in philosophy. This was a theory of the origin of living forms, and it stemmed from a generalization of the inevitable inferences from the exploration of huge masses of facts related to diverse spheres of biology. One should also account for the fact that, in the XIXth century, Darwinist theory still remained a hypothesis hardly verified by contemporary experimentation. The core of evolutionary theory was the fact that Darwin had revealed the objective causal relations between natural phenomena; he also demonstrated the contradictions which produce the contents of the development process and act as its stimuli. Thus Darwin provided a causal representation of the evolutionary process as a form of biological necessity.

As a result, positivism inevitably greatly hindered dissemination of the evolutionary theory in the West and came into conflict with this theory, as a result of which positivism failed altogether. Meanwhile, evolutionary theory – which relies on the methodological principles of spontaneous dialectics – contributed to the further progress of materialism and to its consolidation. Subsequent promotion of these relations gave rise, in the twenties of the XXth century, to a group of evolutionary-minded philosophers who were greatly influenced by F. Engels' *Dialectics of Nature.*

In the history of the experimental method one can distinguish a well-defined period of its flourishing, truly a triumph, when it became widely appreciated in embryology. At that time the experimentally-minded biology of early development closely cooperated with physics, chemistry, and medicine. In other words, it turned into a comprehensive science. It was during this period that at last it became possible to undertake research in isolated blastomers and egg fragments. It was at this time that experiments carried out by Hans

Driesch caused the emergence of neovitalism in experimental embryology. Experiments performed by this science provided the basis of facts upon which there appeared generalizations of philosophy which finally brought Driesch to a vitalistic interpretation of the processes of life. In developing a new methodology, neovitalists failed to explain numerous phenomena and facts by mechanistic argumentation and proved incapable of revealing their causal relations.

The neovitalists' major mistake was rooted in the fact that they tried to explain the recent discoveries made by experimental biology, physics, and chemistry from an idealistic angle. Neovitalists left unsolved and unexplained numerous problems involved in applying a rapidly developing experimental method. A sharp decline in neovitalism was observed in the thirties of this century, when the achievements of genetics and biochemistry brought biology much closer to the understanding of the mechanisms operative in heredity and growth.

Yet, philosophical discourse around the ideas of 'vital force', 'entelechy', 'dominant', etc. promoted research in these trends in biology. Biology, as well as philosophy, could be supplied with a truly objective, materialistic interpretation of the mechanism of heredity only after dialectical materialism was finally accepted as the methodological foundation in the above-mentioned branches of science.

E. M. MIRSKY

PHILOSOPHY OF SCIENCE, HISTORY OF SCIENCE, AND SCIENCE OF SCIENCE

1. The central topic of today's meeting, cooperation between philosophy of science and history of science, is, to my mind, primarily a problem of professional interaction between philosophers and historians of science in studying some common object. It appears reasonable to emphasize this aspect to bring to the fore the research nature of the interaction under consideration. As far as its contents and its typological characteristics are concerned, we should start with analysing already existing forms of interaction between history and philosophy. Such an analysis might provide us with an evaluation of the accumulated experience, enable us to reveal the specific features of our subject, and prevent digressions along the lines which have already proved inefficient.

In this context, development of philosophy itself is a most interesting example of interaction between research in theory and research in history in an analysis of cognitive activities.

There is a specificity characteristic of interaction between theory and history in comprehending the object of philosophy, namely, both types of research have ever been undertaken by the same community (and not infrequently, by the same philosopher), thus being just different forms in which the growth of some general sum of knowledge took place. The form of development in which this sum of knowledge grew, critical reflection, was uniform, at least in the European tradition.

2. From some of the communications offered for discussion, one can infer that the scheme of interaction described by the authors appears to them optimal, if not exclusive; also the interaction between philosophy and history of science, despite all difficulties encountered in implementing such interaction.

To my mind, attempts at consistent analogy between a philosophical research (theoretical or historical) and that in philosophy and history of science cause hindrances of a fundamental nature—and these are to be revealed at the earliest possible stages of the inter-

J. Hintikka, D. Gruender, and E. Agazzi (eds.), Pisa Conference Proceedings, Vol. II, 295–299.

action under discussion. In the first place, I refer to the problem of the historian's or philosopher's attitude towards his object of study. To be more precise, we should understand what kind of ideas concerning science he is to analyse, and how he is to accept these ideas.

From this angle, too, it seems quite reasonable to expect a historian or philosopher of science to acquire a professional standard in the understanding of the problems raised by some specific branch of science, such as physics or chemistry, that he should have accumulated some experience in the field and be able to evaluate its situation "from the inside". Methodologically, this requirement is far from being above criticism. There are some considerations beyond the fact that a scientist can hardly become a well-qualified professional in some specific science, philosophy of science and history of science. It would be naive to demand that one should become professional in every branch of science in order to be allowed to voice his judgements of science as a whole, or to expect that one would become professional in 18th century research in physics (including some practical experience). Our objections deal with some more general aspects involved in the treatment of science as an object of study in modern philosophy of science and history of science.

3. As far as analysis of science was understood as analyses of some selected scientific problem or theory, isolated from its scientific environment, there was no point in discussing interaction between history and philosophy of science – beyond meagre attempts to apply some general statements of philosophy of history to history of science. A philosophical research deals with the structure of the exemplar theory or the links between this issue and dominant ideas of cognition, style of thinking, world outlook, and so forth. On the contrary, history of science traces temporal progress of a *single* problem. At best, discussions were restricted to the ways by which some extra-scientific factors (social, cultural, psychological, *etc.*) should be taken into account or eliminated in the development of a *single* isolated problem. In this sense, the interest revealed by philosophy to the dynamic aspects, to the progress of a single theory, could not change anything basically. For the philosopher, history of science is no more than a chronological presentation of facts as such, and the philosopher has to reveal the dynamic interaction between them (the "internal history") in the course of a "rational reconstruction" ignoring the conceptions set forth by

historians. In his turn the historian of science is far from being certain of what he is to do with such a "rational reconstruction", devoid of any historical content – for him it tries to prove that the facts under consideration are quite lawfully arranged in the sequence in which the historian found them.

4. It was only recently that detailed and profound discussion covered the ways in which the *intra*scientific context affects the formulation, structure, and development of an exemplar problem or theory: in other words, how they are affected by investigation of some other problems and the progress of knowledge in a scientific discipline as a whole. The stimulus to this discussion was provided by the exchange of opinions on Kuhn's book. Some aspects of this discussion are significant for our topic.

The central of these is misidentification of the objects of a philosophical and an historical investigation of science. According to Kuhn, progress of science is the result of efforts applied by the members of a community to *different* problems of a disipline (for Kuhn, it is primarily physical optics). Such a *sum of problems* basically cannot be structured as a uniform theory, and coordination of these problems is seen in the light of a much less definite and much more amorphous idea of a paradigm.

It is worth mentioning that within the decade which has elapsed since the discussion on the growth of knowledge based on Kuhn's book, all the attempts at specifying the empirical contents of the concept 'paradigm' demonstrated outwardly an unexpected but, in fact, quite logical deficiency of both history and philosophy of science in the *systematic description* of objects which stand above an individual problem or theory. Meanwhile, the necessary prerequisite of describing a *sum* of interrelated problems is to be seen in the possibility of their integration within a totality at a higher level: of their classification based on some common factor and of carrying out their comparative analysis. It became clear that neither the source of empirical data (it is far from accidental that, in the history of science, study of the sources is much inferior to that in other historical disiplines), nor the method of selecting and evaluating facts, to say nothing of the existing means of their organization, are adaptable to such work, which involves not only intuition but also some objectivized criteria which serve for reference.

5. Such a systematic investigation of science is, at present, prac-

tically widespread. In some countries, including the USSR, such work comes under science of science, in other countries it poses as sociology of science, information science, *etc.*, but in all these cases a comprehensive approach to the study of science is practiced, be it relations in a scientific community, structure and development of research subject matter, information flows, *etc.* Of late, studies of a scientific discipline as a system in its structural and dynamic stability comes to the fore as one of the central objects of such research. As this work widely uses the results of empirical investigations, statements of problems frequently lack generality, and yield to those of traditional problems of philosophy of science in their academic esteem. Still, already now we can mention, in this context, some progress and certain results which are of interest both for philosophical and historical interpretation.

Under this heading fall papers on empirical selection, quantitative and chronological registration of such structural components of a scientific discipline as research directions and fields, research teams and scientific schools, *etc.* What matters here is not agreement on definitions of terms but separation and description of objects in which their structure, forms of behaviour and dynamics admit observation and comparative study. As an example, the results of such studies testify to the fact that a scientific discipline is an hierarchy of conglomerates which differ in size, "life-time" and dynamic rates, which cannot be usefully described by a single term, be it a 'paradigm' or 'community'.

Meanwhile, in the overwhelming majority of cases, empirical facts are obtained from the standpoint of an external observer, whereas those data are indirect in nature. Moreover, experience shows that application of the ideas of those scientists who see the problems 'from the inside' can be useful, provided special work has been done to choose from their statements what can be well regarded as 'facts'. This statement fully applies to the subject matter of history of science. The information accumulated by it forms a large proportion of the materials used by the science of science, and in some cases, makes the entire sum of the data. Yet, these materials cannot be awarded the status of facts before much elaborate processing, since, in their original form, information from the history of science is arranged to solve different problems and cannot be treated as facts for science of science.

In this context, I should like to mention a number of methodological and substantive problems, in the solution of which professional cooperation between historians and philosophers of science seems to me promising as related to research in science of science and its results.

6. The first of these is construction of satisfactory models to classify disciplines, trends, theories, problems, hypotheses, *etc.*, together with disclosure of possibilities for their hierarchical presentation, comparison, and arrangement. Without the above, one cannot be serious in talking about the structure of knowledge within a discipline. As far as diachrony is concerned, the same applies to the problem of periodization in the history of science as a whole, in the history of an individual science, or its components. The principles of temporal periodization by century, type of world outlook, or style of thinking which are now in use are far from being adequate in all cases. In this connection, some attempts are being made, currently, to introduce a unit of periodization, namely, a "scientific generation", though these attempts encounter considerable resistance . In other words, it is the issue of developing some set of analytic units for research in science. To my mind, there is another related problem which is of interest in the philosophical aspect: that of how differentiation within a discipline should be treated, depending on whether we deal with its discontinuation as a set of problems, or its further development, but within the framework of a higher level totality.

In conclusion, I should like to emphasize the main idea once more. Even though *independent* development of philosophy of science and history of science appears quite promising, based on the traditions characteristic of each of these disciplines, their regular and fruitful interaction is possible only provided they apply analyses and explanation of classified information on science accumulated by science of science and deal with problems of obtaining such information.

L. I. UVAROVA

INTERRELATIONS BETWEEN HISTORY OF SCIENCE AND PHILOSOPHY OF SCIENCE IN RESEARCH IN THE DEVELOPMENT OF TECHNICAL SCIENCES

Prof. E. Aggazzi's report presents some ideas on the evolution in philosophy of science and history of science. He is quite justified in noting that the general trend of their progress points towards close interaction between these branches of knowledge.

Current research in some fields of history of science may be regarded as proof of beneficial influences caused by philosophy of science in the treatment of original and basic aspects of a problem in history of science. In this context, I should like to draw your attention to history of technical sciences as a specific branch of history of science.

It is by far less frequent that history of technical sciences, rather than history of natural sciences, comes to the focus of attention among philosophers of science; one can well say that philosophy of technical sciences has not yet outgrown its embryo phase. However, progress in the philosophy of these sciences is a critical issue both for the growth of philosophy of science in general, and history of technical sciences in particular.

A researcher in the history of technical sciences faces considerable obstacles caused by numerous factors. The empirical material which he is to examine carefully while studying progress in technical sciences is to be found in numerous and diversified sources. Information contained in these sources varies widely in its nature, language (descriptive texts, calculations, graphic illustrations, models, actual structures, etc.), and in time – from ancient times to the present day. Besides, one should account for some specificities of technical sciences which significantly hinder the exploration of their progress. Among these, mention should be made, first of all, of the close ties with practical activities, correlation between cognitive and generative functions, applicability to design, utmost vulnerability to social fac-

J. Hintikka, D. Gruender, and E. Agazzi (eds.), Pisa Conference Proceedings, Vol. II, 301–304.

tors, etc., which, in sum, result in very high development rates. To a great extent, these difficulties encountered in approaching technical sciences as an object of a study in history are responsible for the fact that history of technical sciences is known much worse than that of natural sciences.

If we approach the history of technical sciences in a most general way, we can subdivide it into two basically different periods. The first, which covers emergence of technical sciences, is, in a sense, similar to history of natural sciences. In spite of the fact that the object of natural sciences is nature as such, whereas that of technical sciences is man-made environment, they both started with studying an already existing object. To some extent, research in history of technical sciences during that period could rely on cooperation with philosophy of natural sciences, from which it borrowed some methodological concepts common to them due to to the above-mentioned similarity.

The second period, that of well-developed technical sciences, is basically different from the first. During this period, technical sciences were active in shaping the objects of their investigation, and that resulted in the emergence of a peculiar system characterized by intense feedback. Jointly with some other features characteristic of technical sciences, this factor caused such extreme difficulties in the study of their development that, in all probability, no adequate model representing the history of the technical sciences could be produced without some new principles of methodology and new concepts.

Here, again, historians and philosophers of science should refer to the history of research in natural sciences. To be more precise, with a view to the fact that in research in natural sciences two relatively separate trends in science have taken shape and, later, the need for their closer cooperation resulting in mutual enrichment was duly recognized, one should admit that in the technical sciences, the history and philosophy of which are still very young, extreme care should be taken to eliminate isolationism. The recognition of the great value of a due interaction between the history of the technical sciences and the philosophy of these sciences may become highly promising in promoting accelerated growth in newly emerging fields of science.

Progress in the technical sciences must be studied jointly with the philosophy of the technical sciences. Their interaction may proceed along two lines: first, as a basis to arrange research teams; second, as a direction in the activities of an individual scientist.

Researchers in the history of the technical sciences are entitled to expect that such efforts jointly with the philosophy of these sciences will contribute to the shaping up of methodological approaches and ideas which will primarily help formulate fundamental concepts which still remain basically undefined (such as 'technical sciences', 'materiel', 'technology', 'technical knowledge', 'invention', etc.). Besides, they hope that this cooperation will facilitate development of principles underlying chronological classification for the history of the technical sciences, help disclose the ways of interaction between the technical, natural, and social sciences, define the role played by technical sciences in the 'science-production' system, understand the place of technical sciences in transforming science into an immediate production force of society, etc.

In conclusion, we should like to mention one more fact relevant to the history of the technical sciences. The above-mentioned specific feature of the technical sciences, the shaping of their own object of study, has an important corollary: methodologically well-grounded history of technical sciences is capable of reaching the high standards of applied research which, among other functions, may be useful in producing scientific and technical prognoses.

By analysing development stages in technical sciences, by studying peculiarities characteristic of scientific and technical activities at each stage, by revealing relations between technical sciences, natural science, social sciences, and facts of social life the history of the technical sciences will grow beyond the stage which still restricts it to the accumulation and description of facts. It will also raise its standards to form a structure of knowledge capable of performing the prognostic function.

BIBLIOGRAPHY

Agassi, J.: *Science in Flux*, D. Reidel, Dordrecht, 1975. 12. 'The confusion between science and technology in the standard philosophies of science', pp. 282–305. 13 'Positive evidence in science and technology', pp. 306–321. 14. 'Positive evidence as a social institution', pp. 322–337.

Albraght, E., and Kant, H.: 'Philosophische problems des Eyklus wissenschaft-Technik-Production', *Deutsche Z. Für Philosophie*, B., 1976, Jg. 24, N3, S. 245–260.

Banse, G.: 'Philosophische Fragen der technischen Wissenschaften – Probleme und

Ergebnisse', *Deutsche Z. für Philosophie*, B., 1976, Jg. 24. N3, S. 307–318.
Buchanen, E.: *Technology and History.*
Buchanan, R.A.: The Promethean revolution: Science, technology and history, in: *History of Technology*, I ann. vol. 1976, L., 1976, pp. 73–83.
Friedrich, H.: 'Philosophische Fragen der Technologieproblematik', *Deutsche Z. für Philosophie*, B., 1977, Jg. 25, N1, S. 5–14.
Daumas, M.: 'The history of technology: its aims limits, its methods', in: *History of Technology*, I ann. vol., 1976, L., 1976, pp. 85–112.
Jarvie J.: 'The social character of technological problems', in (ed.), *Contributions to a Philosophy of Technology*, Dordrecht–Boston. 1974, pp. 86–92.
Pacey, A.: *The Maze of Ingenuity: Ideas and Idealism in the Development of Technology, L.*, 1974.
Philosophical Problems of Science & Technology, ed. by A.C. Michalos, Boston (Mass.), Allan & Bacon, 1974, XII, 623 p., Ind: pp. 605–623.
Rapp, F.: 'Technological and scientific knowledge', in: 5th *International Congress of Logic, Methodology and Philosophy of Science.* L., 1975. Contributed papers. s.1 (1975).
Skolimosky, G.: 'Structure of technological thought', in: *Contribution to a Philosophy of Technology*, Dordrecht–Boston, 1974, pp. 72–85.
Melestschenko, 'U.S. Technical sciences', in his book: *Technique and regularity of its development*, Leningrad, 1970 (in Russian).
Uvarova, L.I.: *Scientific Progress and Development of Technical Means*, Moscow, 1973 (in Russian).
Specification of Technical Sciences, Moscow, 1975 (in Russian).
Ivanov, B.I. and Cheschev, V.V.: *Formation and Development of Technical Sciences.* Leningrad, 1977, (in Russian).

M. S. BASTRAKOVA

FROM HISTORY OF SCIENCE TO THEORY OF SCIENCE: AN ESSAY ON V. I. VERNADSKY'S WORK (1863–1945)

"Round table" discussions focus around cooperation between historians of science and philosophers in the exploration of science as a phenomenon. A successful solution to this problem will, in the end, not only benefit progress in history of science but also facilitate settlement of major issues involved in the general theory of scientific knowledge.

How can history of science and philosophy benefit each other? In seeking the answer to this question it may be useful to refer to some forgotten or little-known pages of the history and philosophy of science. In this context, a study of V. I. Vernadsky's works in the history of science may be of great value. This scientist belonged to those naturalists who already, as early as at the rise of the 20th century, became fully aware of the urgency of promoting the philosophy of natural science and of studying its history as an integral unity. The major point is that he was able to comprehend the role of a philosophical interpretation of scientific growth for the shaping of the theory of science. His papers contain formulations of a number of important problems of the history of science; moreover, he managed to outline the ways towards their solution.

V. I. Vernadsky is recorded in the history of science as a geochemist, whereas his creations in the history of natural sciences are relatively little known. His heritage in this field, mostly found in unpublished manuscripts, is, in fact, only being approached with a view to investigation and further development. V. I. Vernadsky's papers on the history of science were written between 1902 and 1938. The latest of these, *Scientific Thought as a Planetary Phenomenon*, is a kind of generalization covering the results of research of many years in geochemistry, history of science, and philosophy of natural sciences, the author's summary of deliberations on the nature of science, the laws governing its progress, the mechanisms by which it

J. Hintikka, D. Gruender, and E. Agazzi (eds.), Pisa Conference Proceedings, Vol. II, 305–309.

is related to its environment, and on its logic and methodology. It is in this paper that his approach to history and, what is of utmost importance, his original positions in theory and methodology are presented.

The major and most characteristic feature of V. I. Vernadsky as a historian is involvement in research in the history of natural science in close coordination with theory of science, which is an inherent feature of his approach. The scientist's starting point can be formulated in his own words as follows: "science is inseparable from philosophy and cannot grow in the absence of a philosophy" [1]. This idea permeates each of his works. It is noteworthy that the idea of inseparable links between science and philosophy, which was of prime value in V. I. Vernadsky's creative work, was rooted in a specific analysis of the history of natural science. According to him, it is the history of scientific thought that makes evident the impact of philosophy "as the source and air breathed by scientific searches" [2]. In 1902, in the preface to his first paper on research in history, 'Development of modern scientific world outlook', he wrote: "Studying the history of scientific thought we see that philosophy's ideas and conceptions have ever been an inalienable and all-permeating element of science since it came to life" [3].

V. I. Vernadsky believed in close contacts between philosophers and historians of natural science, who will exchange their results and methods, and criticize each other. He paid much attention to their joint efforts in investigations of boundary problems of mutual interest. He felt the problem of progress in knowledge to be one of these, and was sure that work in this direction must be carried on with mandatory participation of philosophers. In 1921, V. I. Vernadsky set forth a proposal on the setting up of a special Commission on the History of Knowledge which was to coordinate comprehensive research in the history of science, history of technology, and philosophy. In 1939, he again raised the issue of joint effort by historians of science and philosophers. This idea is expressed in his draft statute prepared for the first Soviet Institute of History of Science. He argued to the effect that the need for their close cooperation in studying the course of progress in scientific thought was rooted in the specificities of the object of study, which is a comprehensive one in its very nature.

There are some other aspects of the philosophy of natural science which also deserve a similar approach. In the thirties, V. I. Vernadsky investigated interrelations between space and time, and he paid much attention to the historical and philosophical aspects of the problem. He gradually arrived at the conclusion that the investigation of such an intricate matter "should involve permanent contacts between philosophers, psychologists, historians, and physicists, rather that physicists alone". Such contacts were to "keep philosophical thought at a required standard of knowledge", on the one hand, and secure a scientific atmosphere in which scientists and historians might "successfully select facts of science and construct their working hypotheses" [4].

Generally speaking, V. I. Vernadsky saw cooperation and mutual assistance between historians of science and philosophers as follows: philosophers apply their long-verified techniques to provide logical profoundness to the research process, while historians of science, in their turn, provide philosophers with original facts to be analyzed, and with a tool with which to evaluate the results obtained—this tool being the method of criticizing a scientific truth from the historical angle.

In his works, V. I. Vernadsky repeatedly emphasized the role of the history of science as an instrument of research in the most general aspects of natural sciences. Once and again, he stressed the significance of comprehending the all accepted scientific and philosophical concepts to understand the essence of modern world outlook, rather than its sources alone. Investigation into the ways of scientific knowledge helps one "reconstruct the historic perspective, comprehend the actual impact of scientific ideas, notions and constructs". By analyzing and evaluating the present situation, the history of science paves the way to further progress in scientific thought, and thus it becomes a "means of reaching new heights" [5].

The revolution in natural science played a major role in shaping V. I. Vernadsky's ideas on the history of science and philosophy, as had been the case with his elder contemporaries, K. A. Timiryazev, W. Ostwald, H. Poincaré, who also dealt with investigations in science as such, and creative scientific activities. Like the above scientists, V. I. Vernadsky believed in the promotion of history of science based on formulation of the most urgent aspects of its general theory. Already as early as at the dawn of his creative work, he wrote:

Historically-oriented investigation of science is to aim at disclosing the laws governing progress of thought, emergence of geniuses, at investigating conditions under which discoveries are made, studying the inherent ways of the methods of scientific thinking, experimentation and observation [6].

In the twenties and thirties, the range of problems to be historically analyzed broadened to include some aspects of relations between science and the State, science and philosophy, the role of creative scientists and the "scientific environment", "the sources of the current revolution in science", the shaping up of the structure of modern scientific knowledge. It can be seen that a number of very important problems of general theory were raised, and many of these still attract researchers in our days. It was the history of science that was expected to contribute heavily to the solution of these problems.

V. I. Vernadsky's papers, not only in the history of science but also of a specific nature, provide a good example of a historically-oriented approach to analyses of some problems raised by the philosophy of natural science. The scientist builds his research upon a solid foundation of empirical facts supplied by the history of scientific knowledge. He proceeds from the history of natural science to its theory.

Studies in the evolution of the biosphere, history of mankind, and progress of scientific thought enabled V. I. Vernadsky to formulate his concept of the nature of science and scientific progress. He believed that science is not something "abstract and self-sufficient" but part of "a complex natural and social structure", of a comprehensive system in which all elements, including the planet with everything living on it, the intellect and scientific thought proper, continuously grow and interact with each other [7].

V. I. Vernadsky's book, *Scientific Thought as a Planetary Phenomenon*, in which he proposed his explanation of the genesis and development of science, was written in 1938, at a time when A. Koyré's and R. Merton's works in history which, according to current evaluations, initiated two major methodological trends in the historiography of science, first saw light [8]. To V. I. Vernadsky, there was no problem of "external" or "internal" factors in the progress of scientific knowledge.

In order to understand the current progress of science one should first of all place it within the framework of a scientific coverage of reality ... Science cannot admit the existence of facts independent of reality, since it proceeds from the unity of reality [9].

According to V. I. Vernadsky, the evolution of the biosphere and the development of "socially organized mankind" are the general factors which determined the emergence of science and continue to determine its further progress [10].

Another important generalization made by V. I. Vernadsky in connection with analyses of progress in scientific knowledge deserves mentioning. This is his well-known thesis to the effect that science is becoming a "geological force" which is active in transforming the planet and preparing a new phase in its evolution: transformation of the biosphere into the noosphere [11].

In V. I. Vernadsky's creative activities, interaction between the history of natural science and the philosophy of natural science was profoundly organic and inherent. His philosophical inferences naturally appeared as a result of his research in history.

REFERENCES

[1] Vernadsky, V. I.: *Scientific Thought as a Planetary Phenomenon.-Deliberations of a Natural Scientist*, Book II, Moscow, 1977, p. 137 (in Russian).

[2] Vernadsky, V. I.: Letters. USSR Academy of Sciences Archives, file 518, op. 71 No. 48, sheet 54.

[3] Vernadsky, V. I. 1922: *On Scientific World Outlook. Essays and Speeches*, Petrograd, vol. 2 (in Russian), p. 20.

[4] Vernadsky, V. I. 1975: *On Life (Biological) Time-Deliberations of a Naturalist*, Book 1, Moscow, p. 112 (in Russian).

[5] Vernadsky, V. I. 1926: 'A note on the urgency of resuming work in the history of sciences', *Proceedings of the General Meetings of the Russian Academy of Sciences*, Leningrad, para. 93 (in Russian).

[6] Vernadsky, V. I.: Letters, USSR Academy of Sciences Archives, file 518, op. 7, No. 40, sheet 48.

[7] Vernadsky, V. I.: *Scientific Thought as a Planetary Phenomenon*, pp. 37-39, 91, 96.

[8] Mikulinsky, S. R. 1971, and L. A. Markova: *Major Trends in the Methodology of History of Science Abroad*, Moscow, pp. 15–17.

[9] Vernadsky, V. I.: *Scientific Thought as a Planetary Phenomenon*, p. 37.

[10] Ibid, pp. 31-33, 43-44.

[11] Ibid, pp. 19–22, 33.

JOHN E. MURDOCH

UTILITY *VERSUS* TRUTH:

At Least One Reflection on the Importance of the Philosophy of Science for the History of Science

There is little doubt that the more appreciative, and thus better rewarded, partners in the controversial "marriage" of the history and philosophy of science are the philosophers. At one recent meeting that I was privileged to attend, a meeting expressly designed to recognize, if not to celebrate, that "marriage," one historian's recognition of the *raison d'être* of the meeting took the form of asking for divorce. And that is not merely a sometime thing. For several years ago, Michael Crowe canvassed American historians of science concerning their annual meetings and the result was that that group with whom they would *least* like to meet was the philosophers of science! That surely is very far from a state devoutly to be wished. What is to be done? Frankly, I do not know. But I believe I do know at least a part of what is to be, or might be, said.

In saying it I shall be thinking only of what the history of science can gain from the philosophy of science (and not vice versa, since my historical colleagues are in greater need of saving than are my philosophical ones). Secondly, I shall be speaking as an historian of early science, as a "pre-Galilean man." And lastly, I shall try to introduce that modicum of exaggeration that is appropriate and needed for a round-table discussion.

The aversion of historians of science for their sister discipline comes in various guises and degrees. Of course, I do not wish to maintain that *all* historians of science suffer from such an aversion, since there are clearly saints among us. But saints are by their nature far and few between and, besides, for many who might like to try the path to sainthood, it is not even clear what the process of canonization might be. What is more, the best way to approach the question of what the philosophy of science can do for the history of science is to concentrate our attention, not on the saintly few, but on the unsaintly majority. There the first level of aversion is usually expressed in a

J. Hintikka, D. Gruender, and E. Agazzi (eds.), Pisa Conference Proceedings, Vol. II, 311–319.

form something like: "Philosophers of science never get the history right." Or: "Philosophers of science may write history as it should have–or perhaps could have–been, not as it really was." The next level of disturbance has to do with the fact that many historians of science actually fear the philosophy of science. True, they can understand and hence brave the likes of a hypothetico-deductive model, but they seem to flee with horror from anything like a propositional function or a Ramsey-sentence. And the final, almost apoplectic, stage of aversion finds expression in the belief that what philosophers of science do is often simply not real (that is, not historical), and even "silly."

Now these last two levels of aversion seem to be due to the simple fact that the historians of science in question just do not understand what they are so opposed to, and I think it rather unlikely that much can be done to cure them of this lack of understanding. The prospects of "cure" are better, however, for those historians whose major complaint with the philosophy of science is that it encourages and ends up with "history as it should have been" not "history as it really was." The answer to such a complaint is really not difficult to formulate. It begins from the fact that it is perfectly all right that the philosophy of science may often–indeed, may even always–end up with "history as it should have been." But this in no way means that what the philosopher of science has accomplished is not valuable, even very valuable, to the historian of science. I would urge that not only *is* it almost always valuable, but that it is valuable on *historical* grounds. That is to say, if an historian could be made to appreciate and to utilize properly the "should have been history" of the philosopher, then this use just might lead that historian to the doing of some "history as it was" of his own, some "history as it was" that would not have been done, perhaps could not have been done, without the "should have been" stimulus and tool of the philosopher.

I shall try to explain at least the structure of what I have in mind in a moment, but before that it seems worthwhile to direct our attention to two other matters. The first is that historians of science have traditionally not shown aversion or impatience when things have been distorted, not by the philosophy of science, but by science itself. The utilization of modern science and mathematics in doing the history of science (and I have in mind especially eras earlier than that of the nineteenth century) has not only been regarded as permissible, but often as the most appropriate way to carry forth this historical enterprise.

One would, for example, put the deliberations of Aristotle on local motion into modern mathematical notation without even asking whether Aristotle had the establishment of any mathematical relation in mind in the first place;[1] or one would systematically translate earlier mathematics into the contemporary way of expressing things without much concern over whether it was proper to do so.[2] To be sure, some historians have seen the dangers of such procedures (one thinks, for instance, of E. J. Dijksterhuis's invention of a new notation to deal with Greek mathematics);[3] but the predominant reaction has been not to react at all, the verdict about "translating old science into new" being essentially one of the resounding approval.

Indeed, certain segments of the history of science have been extraordinarily subject to being viewed through the filter of modern scientific conceptions and developments, one of the most notable cases in point being, I believe, the history of ancient atomism. Thus, what was long one of the most influential books in English on this topic lauded Greek atomism as superior to other scientific "systems" of antiquity because the author viewed (wrongly) this particular development in fifth-century B.C. natural philosophy in terms of nineteenth-century mechanistic determinism.[4] But this kind of misinterpretation and distortion seems to have been endemic wherever atoms or atomism have reared their heads in history. The existence of modern counterparts have made it too much to resist. In fact, wherever respectable "modern scientific entities" have been seen to exist in the earlier history of science, more often than not they have been excised from their context and, in their delightful isolation, touted as what was of most importance in the scientific development at hand. One can think, for example, of the relevance of commentaries on the *Book of Genesis* for the history of science and how very much more attention has been paid to the likes of birds and to the puzzle of the so-called supra-celestial waters because they could be related to cosmological and natural entities than was paid to interpretations of the initially unformed and void state of the earth which did not have the good fortune of such "connections."[5] The irony in this is, in part, that the former really led to nothing much at all of subsequent value in the history of natural philosophy, while the latter was involved in the development of a general theory of generation of physical objects which played an important role in subsequent natural philosophy.[6] Alternatively and more broadly, a great deal – indeed the major portion – of the whole history of earlier medicine has so concentrated its gaze upon the new being recognizable in the old (didn't they have anesthetics in the Middle Ages?, wasn't this or that disease or drug this or that in modern terms?, *etc.*) that an adequate history of medical theory during the period has never been written.

I have drawn attention to this tendency of historians of science to utilize science itself as an interpretive tool in plying their discipline not because I believe that the results (distortions of whatever sort or to whatever extent included) have not been of value. They have been; we have learned much from them. Rather, I have done so primarily to point out that historians of science have not been disturbed when one body of conceptions and doctrines have been brought to bear in doing the history of science and have, as a result, often come up with "history as it was not." But if one is at ease when this body is science itself, why should one be perturbed when it is instead the philosophy of science? The answer is simple: one should not be. In point of fact, I believe that it can be argued that it is potentially *more* dangerous to doing good history of science when science itself is one's analytical tool than when the philosophy of science serves that role. For historians have been far less aware, and are perhaps bound to remain less aware, of the possible distortions introduced by using science than of those distortions subsequent to the utilization of the philosophy of science. Yet since the possiblities of distortion are clearly there in both instances, an equality of awareness relative to such possibilities is the least that one should expect.

The second matter that requires note before I attempt to delineate my own view of the value of the philosophy of science for the history of science is to address the question of why it may be that more historians of science do not make use of the efforts of philosophers and hence do not reap gains, I would claim, for their own ends. One answer to this is, I think, basically because historians are far too much concerned with whether what the philosopher says is *true*, where the truth in question has to do with whether "science really works like that" and not simply with whether the philosopher's claims somehow yields "true history." Correspondingly, historians are less worried about the application of elements of modern science in doing history because they are not troubled by the "problem of truth" in this case; they simply know that it is true or right. Yet when it comes to that less canonized animal called the philosophy of science, at times there is a suspicion, perhaps even a fear, among historians that if, in doing history, one makes use of what the philosopher has maintained, then one is somehow philosophically committed to the rightness or truth of what the philosopher has held. But is not the fact of the matter that the historian can very well *use* the philosopher's contentions without in any way being committed to their truths?

The reply to this might read: "well, perhaps so, but the contentions of the philosopher really do not *fit* what I find to be the case historically, so why should I wish to utilize anything like that in doing history?" Yet, while drawing attention to a factor of crucial importance, such a reply really misses the point. For the real value for the history of science of the contentions of the philosopher of science can be seen to derive from the very fact that these contentions almost never do fit what was historically so. (Indeed, one can remove the 'almost' and simply say "never," since if one looks long and hard enough, some lack of fit is always there to be discovered.) The historical value of such an "ill-fit" arises because it can cause the historian to become aware of features in the historical situation to which the philosophical doctrine is applied that one might not be aware of without the "ill-fit" generated by this application. And I do not mean only the ill-fit of the philosophies of the history of science formulated by Kuhn, Lakatos, Toulmin and their *confreres*, or even only the ill-fit arising from conceptions or doctrines germane to philosophy of science in general, but the ill-fit resulting from the application of *any* philosophical conception or doctrine. In a word, these conceptions or doctrines can function as valuable tools of historical analysis precisely because they do generate such "ill-fits."

Let me give several examples of the type of phenomenon I have in mind. In 1930 Heinrich Scholz published an article entitled 'Die Axiomatik der Alten,' its burden being an axiomatic treatment of major segments of Aristotle's *Posterior Analytics*.[7] Now I think it is fair to claim that Scholz's axiomatizing did not exactly fit what Aristotle maintained in the *Posterior Analytics*. Aristotle does not have, for example, postulates of reference, evidence, or deducibility in *precisely the same sense* (a sense drawn from modern logic) as they are taken by Scholz. Yet it surely does not follow that what Scholz had accomplished should be rejected out of hand as mere "should have been history." Rather, does it not mean that, given the incomplete fit of Scholz's claims, one just might be in a better position to elucidate exactly what it is that Aristotle *does* have? More than that, if it is true that Aristotle does not have precisely what Scholz maintains, can we not go on and ask *why* Aristotle does not have precisely that? One might, for example, conclude that Aristotle *could* not have had this or that claimed by Scholz through realizing that what additional element he would have needed in order to make Scholz fit is an element which was not available to, perhaps not even conceivable to, Aristotle. And if something like that is so, then can one not reasonably maintain that Scholz's contentions have functioned as a tool that has enabled one to accomplish at least some historical analysis that might very well have not been accomplished without it?

The elements of contemporary philosophy that Scholz applied in his analysis of Aristotle and also of other historical scientific works[8] were drawn from recent developments in logic and the philosophy of mathematics, but other segments of twentieth-century philosophy (and not just philosophy of science) can play similar roles as productive historical tools. It would seem, however, that the area of logic and essentially logically-based conceptions and techniques have thus far proved extraordinarily fruitful. It is inconceivable, for example, that the *historical* analysis of Aristotle's treatment of future contingents in chapter 9 of his *De interpretatione* would not have been as provocative and as worthwhile as it has been if it had not had the benefit of notions and doctrines drawn from modern logic.[9] Similarly, I do not think that we would today understand as much as we do about William of Ockham's doctrine of supposition were it not for the fact that attempts were made to analyze it in terms of modern quantification theory on the one hand and Lesniewskian ontology on the other, neither of which really fit what Ockham had in mind.[10] But again, the lack of fit has helped us in working toward, if not yet actually reaching, just what was historically the case. And I would think that similar historical advance will result from such very recent applications of the techniques of modern logic to history as that of Hintikka and Remes to Pappus of Alexandria.[11]

Moving away from the utilization of mathematical logic in doing the history of science, the value of philosophical analysis of a more general sort is strikingly evident in Ernst Mach's criticism of Archimedes's proof of the law of the lever and the historical and philosophical "setting things straight" which this criticism generated. In effect, Mach charged Archimedes with a *petitio* insofar as in deducing the general law of the lever he had tacitly introduced experience containing this general law.[12] Beginning with Otto Hölder and Giovanni Vailati, this charge set into motion a series of counter-critiques of Mach, a good deal of the effort that was then spent being directed toward establishing just what, in matter of historical fact, was Archimedes's procedure.[13]

In some of the above "case histories" of what I have in mind as the appropriate and fruitful use of the philosophy of science in doing the history of science, the philosophy in question may not have been "up to date" (indeed, may have later even received rather definitive philosophical criticism of its own). Yet for the purposes of the historian's use that I am here urging of this philosophy, this makes little difference. The test of its value to the historian is how much any philosophical notion or doctrine can enlighten *historical analysis* (and I mean historical analysis that eventually results in a piece of "history as it really was" which would satisfy the most picayune of historical scholars). The question is one of utility, not of truth. Of course, since this utilization is most likely to issue in slightly skewed history, one must be careful to "unplug" whatever philosophical tools of analysis have been applied once they have served their task of "enlightenment." But this should follow all but automatically, given the fact that what I have called the "ill-fit" generating this enlightenment in the first place already bespeaks one's heightened awareness of the possible presence of historical incongruities.

Such then is at least one fashion in which the philosophy of science – or better philosophy *tout court* – might be of historical value to the historian of science. I realize that the fact that my own concentration in the history of early science, when so much "science" was identical with philosophy, might mean that a greater number of philosophical conceptions and doctrines can be applied as "historical tools," but that certainly does not mean that the function of the philosophy of science need be different in more modern periods of the history of science. And I realize that confining one's use of the philosophy of science to no more than that of an historical tool runs the danger of an historical relativism, but is not this possible disadvantage greatly outweighed by the possible advantage of doing more, and even doing good, history through this use?

NOTES

[1] See, for example, I. E. Drabkin, 'Notes on the laws of motion in Aristotle,' *American Journal of Philology* **59** (1938), 60–84. For an opposing view, see Henri Carteron, *La notion de force dans le système d'Aristote* (Paris, 1923), pp. 1–32.

[2] For example, the "in modern notation" English editions by Sir Thomas Heath of Archimedes and Apollonius. Similarly, the mathematics of proportions applied by Thomas Bradwardine to relating the velocities of bodies in motion to the forces and resistances involved in these motions has usually been explained in modern terms of ratios and exponents; for the proper corrective see J. Murdoch & E. Sylla, 'The science of motion,' in David Lindberg (ed.), *Science in the Middle Ages* (Chicago, 1978), pp. 223–31.

[3] Thus, Dijksterhuis expresses the ratio of A to B as $\Lambda(A, B)$ [Λ for *logos*] or $(AB)^2$ as $T(AB)$ [T for *tetragon*, that is to say "the square on AB"], etc. This first appeared in his *De Elementen van Euclides*, 2 vols. (Groningen, 1929–30), and later in English in his *Archimedes* (New York, 1957).

[4] Cyril Bailey, *The Greek Atomists and Epircurus* (Oxford, 1928). For a partial corrective to Bailey's view: D. M. Balme, 'Greek science and mechanism: II. The atomists,' *Classical Quarterly* **25** (1941), 23–28.

[5] For example: Lynn Thorndike, *A History of Magic and Experimental Science*, vol. 1 (New York, 1923), ch. 21. The treatment of the hexameral literature is more balanced in Pierre Duhém, *Le système du monde*, vol. 2 (Paris, 1914), part 2, ch. 1; and Marshall Clagett, *Greek Science in Antiquity* (New York, 1955), ch. 10, part 2.

[6] I have in mind the originally Stoic notion of *rationes seminales*, its utilization by Gregory of Nyssa and Augustine, its role in the twelfth-century Platonists usually associated with the School of Chartres, and its eventual development within a full-fledged theory of "coming to be" in the thirteenth century in the likes of Bonaventure and Robert Kilwardby.

[7] *Blätter für deutsche Philosophie* **4** (1930–31), 259–78, reprinted in Heinrich Scholz, *Mathesis Universalis* (Basel/Stuttgart, 1961), pp. 27–44.

[8] Heinrich Scholz, 'Die Grundlegung der Euklidischen Geometrie durch Euklid,' *Semesterberichte zur Pflege des Zusammenhanges von Universität und Schule, Münster* **5** (1934), 103–120.

[9] An exhaustive bibliography of recent discussions of Aristotle's treatment of future contingents together with an account of the contents of most of these discussions is now conveniently available in Vincenza Celluprica, *Il capitolo 9 des 'De interpretatione' di Aristotele: Rassegna di studi* 1930–1973, Pubblicazioni del Centro di Storia della Storiografia Filosofica 1 (Rome: Società Editrice il Mulino, 1977).

[10] The literature on this topic is rather extensive, but see: Gareth Mathews, 'Ockham's supposition theory and modern logic,' *Philosophical Review* **73** (1964), 91–99 and

'*Suppositio* and Quantification in Ockham,' *Noûs* 7 (1973), 13–24; D. P. Henry, 'Ockham, *Suppositio*, and Modern Logic,' *Notre Dame Journal of Formal Logic* **5** (1964), 290–92; John Trentman, 'Lesniewski's ontology and some medieval logicians,' *op. cit.* 7 (1966), 361–64; D. P. Henry, 'Lesniewski's ontology and some medieval logicians,' *op. cit.* **10** (1969), 324–26; Graham Priest and Stephen Read, 'The formalization of Ockham's theory of supposition,' *Mind* **86** (1977), 109–113.

[11] J. Hintikka and U. Remes, *The Method of Analysis. Its Geometrical Origin and its General Significance* P. Reidel, (Dordrecht, 1974).

[12] Ernst Mach, *The Science of Mechanics*, transl. T. J. McCormack from the 9th German edition (La Salle, Ill., 1942), pp. 13–17, 19–20, 24–28.

[13] Otto Hölder, *Die Mathematische Methode* (Berlin, 1924), pp. 39–45. Giovanni Vailat's articles are collected in his *Scritti* (Leipzig/Florence, 1911). Mach was able to take account of some of this criticism in the fourth (1901) and fifth (1912) German editions of his work since one of Vailati's articles appeared in 1901, and Hölder had set forth a briefer critique in his *Denken und Anschauung in der Geometrie* (Leipzig, 1900). For further literature on the Mach-Archimedes problem see Marshall Clagett, *The Science of Mechanics in the Middle Ages* (Madison, 1959), p. 12, n. 10.

INDEX OF NAMES

INDEX OF SUBJECTS

SYNTHESE LIBRARY

Studies in Epistemology, Logic, Methodology,
and Philosophy of Science

1. J. M. Bochénski, *A Precis of Mathematical Logic*. 1959.
2. P. L. Guiraud, *Problèmes et méthodes de la statistique linguistique*. 1960.
3. Hans Freudenthal (ed.), *The Concept and the Role of the Model in Mathematics and Natural and Social Sciences*. 1961.
4. Evert W. Beth, *Formal Methods. An Introduction to Symbolic Logic and the Study of Effective Operations in Arithmetic and Logic*. 1962.
5. B. H. Kazemier and D. Vuysje (eds.), *Logic and Language. Studies Dedicated to Professor Rudolf Carnap on the Occasion of His Seventieth Birthday*. 1962.
6. Marx W. Wartofsky (ed.), *Proceedings of the Boston Colloquium for the Philosophy of Science 1961-1962*. Boston Studies in the Philosophy of Science, Volume I. 1963.
7. A. A. Zinov'ev, *Philosophical Problems of Many-Valued Logic*. 1963.
8. Georges Gurvitch, *The Spectrum of Social Time*. 1964.
9. Paul Lorenzen, *Formal Logic*. 1965.
10. Robert S. Cohen and Marx W. Wartofsky (eds.), *In Honor of Philipp Frank*. Boston Studies in the Philosophy of Science, Volume II. 1965.
11. Evert W. Beth, *Mathematical Thought. An Introduction to the Philosophy of Mathematics*. 1965.
12. Evert W. Beth and Jean Piaget, *Mathematical Epistemology and Psychology*. 1966.
13. Guido Küng, *Ontology and the Logistic Analysis of Language. An Enquiry into the Contemporary Views on Universals*. 1967.
14. Robert S. Cohen and Marx W. Wartofsky (eds.), *Proceedings of the Boston Colloquium for the Philosophy of Science 1964-1966. In Memory of Norwood Russell Hanson*. Boston Studies in the Philosophy of Science, Volume III. 1967.
15. C. D. Broad, *Induction, Probability, and Causation. Selected Papers*. 1968.
16. Günther Patzig, *Aristotle's Theory of the Syllogism. A Logical-Philosophical Study of Book A of the Prior Analytics*. 1968.
17. Nicholas Rescher, *Topics in Philosophical Logic*. 1968.
18. Robert S. Cohen and Marx W. Wartofsky (eds.), *Proceedings of the Boston Colloquium for the Philosophy of Science 1966-1968*. Boston Studies in the Philosophy of Science, Volume IV. 1969.

19. Robert S. Cohen and Marx W. Wartofsky (eds.), *Proceedings of the Boston Colloquium for the Philosophy of Science 1966-1968.* Boston Studies in the Philosophy of Science, Volume V. 1969.
20. J. W. Davis, D. J. Hockney, and W. K. Wilson (eds.), *Philosophical Logic.* 1969.
21. D. Davidson and J. Hintikka (eds.), *Words and Objections. Essays on the Work of W. V. Quine.* 1969.
22. Patrick Suppes, *Studies in the Methodology and Foundations of Science. Selected Papers from 1911 to 1969.* 1969.
23. Jaakko Hintikka, *Models for Modalities. Selected Essays.* 1969.
24. Nicholas Rescher *et al.* (eds.), *Essays in Honor of Carl G. Hempel. A Tribute on the Occasion of His Sixty-Fifth Birthday.* 1969.
25. P. V. Tavanec (ed.), *Problems of the Logic of Scientific Knowledge.* 1969.
26. Marshall Swain (ed.), *Induction, Acceptance, and Rational Belief.* 1970.
27. Robert S. Cohen and Raymond J. Seeger (eds.), *Ernst Mach: Physicist and Philosopher.* Boston Studies in the Philosophy of Science, Volume VI. 1970.
28. Jaakko Hintikka and Patrick Suppes, *Information and Inference.* 1970.
29. Karel Lambert, *Philosophical Problems in Logic. Some Recent Developments.* 1970.
30. Rolf A. Eberle, *Nominalistic Systems.* 1970.
31. Paul Weingartner and Gerhard Zecha (eds.), *Induction, Physics, and Ethics.* 1970.
32. Evert W. Beth, *Aspects of Modern Logic.* 1970.
33. Risto Hilpinen (ed.), *Deontic Logic: Introductory and Systematic Readings.* 1971.
34. Jean-Louis Krivine, *Introduction to Axiomatic Set Theory.* 1971.
35. Joseph D. Sneed, *The Logical Sstructure of Mathematical Physics.* 1971.
36. Carl R. Kordig, *The Justification of Scientific Change.* 1971.
37. Milic Capek, *Bergson and Modern Physics.* Boston Studies in the Philosophy of Science, Volume VII. 1971.
38. Norwood Russell Hanson, *What I Do Not Believe, and Other Essays* (ed. by Stephen Toulmin and Harry Woolf). 1971.
39. Roger C. Buck and Robert S. Cohen (eds.), *PSA 1970. In Memory of Rudolf Carnap.* Boston Studies in the Philosophy of Science, Volume VIII. 1971.
40. Donald Davidson and Gilbert Harman (eds.), *Semantics of Natural Language.* 1972.
41. Yehoshua Bar-Hillel (ed.), *Pragmatics of Natural Languages.* 1971.
42. Sören Stenlund, *Combinators, λ-Terms and Proof Theory.* 1972.
43. Martin Strauss, *Modern Physics and Its Philosophy. Selected Papers in the Logic, History, and Philosophy of Science.* 1972.
44. Mario Bunge, *Method, Model and Matter.* 1973.
45. Mario Bunge, *Philosophy of Physics.* 1973.
46. A. A. Zinov'ev, *Foundations of the Logical Theory of Scientific Knowledge (Complex Logic).* (Revised and enlarged English edition with an appendix by G. A. Smirnov, E. A. Sidorenka, A. M. Fedina, and L. A. Bobrova.) Boston Studies in the Philosophy of Science, Volume IX. 1973.
47. Ladislav Tondl, *Scientific Procedures.* Boston Studies in the Philosophy of Science, Volume X. 1973.
48. Norwood Russell Hanson, *Constellations and Conjectures* (ed. by Willard C. Humphreys, Jr.). 1973.

49. K. J. J. Hintikka, J. M. E. Moravcsik, and P. Suppes (eds.), *Approaches to Natural Language.* 1973.
50. Mario Bunge (ed.), *Exact Philosophy – Problems, Tools, and Goals.* 1973.
51. Radu J. Bogdan and Ilkka Niiniluoto (eds.), *Logic, Language, and Probability.* 1973.
52. Glenn Pearce and Patrick Maynard (eds.), *Conceptual Change.* 1973.
53. Ilkka Niiniluoto and Raimo Tuomela, *Theoretical Concepts and Hypothetico-Inductive Inference.* 1973.
54. Roland Fraissé, *Course of Mathematical Logic* – Volume 1: *Relation and Logical Formula.* 1973.
55. Adolf Grünbaum, *Philosophical Problems of Space and Time.* (Second, enlarged edition.) Boston Studies in the Philosophy of Science, Volume XII. 1973.
56. Patrick Suppes (ed.), *Space, Time, and Geometry.* 1973.
57. Hans Kelsen, *Essays in Legal and Moral Philosophy* (selected and introduced by Ota Weinberger). 1973.
58. R. J. Seeger and Robert S. Cohen (eds.), *Philosophical Foundations of Science.* Boston Studies in the Philosophy of Science, Volume XI. 1974.
59. Robert S. Cohen and Marx W. Wartofsky (eds.), *Logical and Epistemological Studies in Contemporary Physics.* Boston Studies in the Philosophy of Science, Volume XIII. 1973.
60. Robert S. Cohen and Marx W. Wartofsky (eds.), *Methodological and Historical Essays in the Natural and Social Sciences. Proceedings of the Boston Colloquium for the Philosophy of Science 1969-1972.* Boston Studies in the Philosophy of Science, Volume XIV. 1974.
61. Robert S. Cohen, J. J. Stachel, and Marx W. Wartofsky (eds.), *For Dirk Struik. Scientific, Historical and Political Essays in Honor of Dirk J. Struik.* Boston Studies in the Philosophy of Science, Volume XV. 1974.
62. Kazimierz Ajdukiewicz, *Pragmatic Logic* (transl. from the Polish by Olgierd Wojtasiewicz). 1974.
63. Sören Stenlund (ed.), *Logical Theory and Semantic Analysis. Essays Dedicated to Stig Kanger on His Fiftieth Birthday.* 1974.
64. Kenneth F. Schaffner and Robert S. Cohen (eds.), *Proceedings of the 1972 Biennial Meeting, Philosophy of Science Association.* Boston Studies in the Philosophy of Science, Volume XX. 1974.
65. Henry E. Kyburg, Jr., *The Logical Foundations of Statistical Inference.* 1974.
66. Marjorie Grene, *The Understanding of Nature. Essays in the Philosophy of Biology.* Boston Studies in the Philosophy of Science, Volume XXIII. 1974.
67. Jan M. Broekman, *Structuralism: Moscow, Prague, Paris.* 1974.
68. Norman Geschwind, *Selected Papers on Language and the Brain.* Boston Studies in the Philosophy of Science, Volume XVI. 1974.
69. Roland Fraissé, *Course of Mathematical Logic* – Volume 2: *Model Theory.* 1974.
70. Andrzej Grzegorczyk, *An Outline of Mathematical Logic. Fundamental Results and Notions Explained with All Details.* 1974.
71. Franz von Kutschera, *Philosophy of Language.* 1975.
72. Juha Manninen and Raimo Tuomela (eds.), *Essays on Explanation and Understanding. Studies in the Foundations of Humanities and Social Sciences.* 1976.

73. Jaakko Hintikka (ed.), *Rudolf Carnap, Logical Empiricist. Materials and Perspectives.* 1975.
74. Milic Capek (ed.), *The Concepts of Space and Time. Their Structure and Their Development.* Boston Studies in the Philosophy of Science, Volume XXII. 1976.
75. Jaakko Hintikka and Unto Remes, *The Method of Analysis. Its Geometrical Origin and Its General Significance.* Boston Studies in the Philosophy of Science, Volume XXV. 1974.
76. John Emery Murdoch and Edith Dudley Sylla, *The Cultural Context of Medieval Learning.* Boston Studies in the Philosophy of Science, Volume XXVI. 1975.
77. Stefan Amsterdamski, *Between Experience and Metaphysics. Philosophical Problems of the Evolution of Science.* Boston Studies in the Philosophy of Science, Volume XXXV. 1975.
78. Patrick Suppes (ed.), *Logic and Probability in Quantum Mechanics.* 1976.
79. Hermann von Helmholtz: *Epistemological Writings. The Paul Hertz/Moritz Schlick Centenary Edition of 1921 with Notes and Commentary by the Editors.* (Newly translated by Malcolm F. Lowe. Edited, with an Introduction and Bibliography, by Robert S. Cohen and Yehuda Elkana.) Boston Studies in the Philosophy of Science, Volume XXXVII. 1977.
80. Joseph Agassi, *Science in Flux.* Boston Studies in the Philosophy of Science, Volume XXVIII. 1975.
81. Sandra G. Harding (ed.), *Can Theories Be Refuted? Essays on the Duhem-Quine Thesis.* 1976.
82. Stefan Nowak, *Methodology of Sociological Research. General Problems.* 1977.
83. Jean Piaget, Jean-Blaise Grize, Alina Szeminska, and Vinh Bang, *Epistemology and Psychology of Functions.* 1977.
84. Marjorie Grene and Everett Mendelsohn (eds.), *Topics in the Philosophy of Biology.* Boston Studies in the Philosophy of Science, Volume XXVII. 1976.
85. E. Fischbein, *The Intuitive Sources of Probabilistic Thinking in Children.* 1975.
86. Ernest W. Adams, *The Logic of Conditionals. An Application of Probability to Deductive Logic.* 1975.
87. Marian Przelecki and Ryszard Wójcicki (eds.), *Twenty-Five Years of Logical Methodology in Poland.* 1977.
88. J. Topolski, *The Methodology of History.* 1976.
89. A. Kasher (ed.), *Language in Focus: Foundations, Methods and Systems. Essays Dedicated to Yehoshua Bar-Hillel.* Boston Studies in the Philosophy of Science, Volume XLIII. 1976.
90. Jaakko Hintikka, *The Intentions of Intentionality and Other New Models for Modalities.* 1975.
91. Wolfgang Stegmüller, *Collected Papers on Epistemology, Philosophy of Science and History of Philosophy.* 2 Volumes. 1977.
92. Dov M. Gabbay, *Investigations in Modal and Tense Logics with Applications to Problems in Philosophy and Linguistics.* 1976.
93. Radu J. Bogdan, *Local Induction.* 1976.
94. Stefan Nowak, *Understanding and Prediction. Essays in the Methodology of Social and Behavioral Theories.* 1976.
95. Peter Mittelstaedt, *Philosophical Problems of Modern Physics.* Boston Studies in the Philosophy of Science, Volume XVIII. 1976.

96. Gerald Holton and William Blanpied (eds.), *Science and Its Public: The Changing Relationship*. Boston Studies in the Philosophy of Science, Volume XXXIII. 1976.
97. Myles Brand and Douglas Walton (eds.), *Action Theory*. 1976.
98. Paul Gochet, *Outline of a Nominalist Theory of Proposition. An Essay in the Theory of Meaning*. 1980. (Forthcoming.)
99. R. S. Cohen, P. K. Feyerabend, and M. W. Wartofsky (eds.), *Essays in Memory of Imre Lakatos*. Boston Studies in the Philosophy of Science, Volume XXXIX. 1976.
100. R. S. Cohen and J. J. Stachel (eds.), *Selected Papers of Léon Rosenfeld*. Boston Studies in the Philosophy of Science, Volume XXI. 1978.
101. R. S. Cohen, C. A. Hooker, A. C. Michalos, and J. W. van Evra (eds.), *PSA 1974: Proceedings of the 1974 Biennial Meeting of the Philosophy of Science Association*. Boston Studies in the Philosophy of Science, Volume XXXII. 1976.
102. Yehuda Fried and Joseph Agassi, *Paranoia: A Study in Diagnosis*. Boston Studies in the Philosophy of Science, Volume L. 1976.
103. Marian Przelecki, Klemens Szaniawski, and Ryszard Wójcicki (eds.), *Formal Methods in the Methodology of Empirical Sciences*. 1976.
104. John M. Vickers, *Belief and Probability*. 1976.
105. Kurt H. Wolff, *Surrender and Catch: Experience and Inquiry Today*. Boston Studies in the Philosophy of Science, Volume LI. 1976.
106. Karel Kosík, *Dialectics of the Concrete*. Boston Studies in the Philosophy of Science, Volume LII. 1976.
107. Nelson Goodman, *The Structure of Appearance*. (Third edition.) Boston Studies in the Philosophy of Science, Volume LIII. 1977.
108. Jerzy Giedymin (ed.), *Kazimierz Ajdukiewicz: The Scientific World-Perspective and Other Essays, 1931-1963*. 1978.
109. Robert L. Causey, *Unity of Science*. 1977.
110. Richard E. Grandy, *Advanced Logic for Applications*. 1977.
111. Robert P. McArthur, *Tense Logic*. 1976.
112. Lars Lindahl, *Position and Change. A Study in Law and Logic*. 1977.
113. Raimo Tuomela, *Dispositions*. 1978.
114 Herbert A. Simon, *Models of Discovery and Other Topics in the Methods of Science*. Boston Studies in the Philosophy of Science, Volume LIV. 1977.
115. Roger D. Rosenkrantz, *Inference, Method and Decision*. 1977.
116. Raimo Tuomela, *Human Action and Its Explanation. A Study on the Philosophical Foundations of Psychology*. 1977.
117. Morris Lazerowitz, *The Language of Philosophy. Freud and Wittgenstein*. Boston Studies in the Philosophy of Science, Volume LV. 1977.
118. Stanislaw Leśniewski, *Collected Works* (ed. by S. J. Surma, J. T. J. Srzednicki, and D. I. Barnett, with an annotated bibliography by V. Frederick Rickey). 1980. (Forthcoming.)
119. Jerzy Pelc, *Semiotics in Poland, 1894-1969*. 1978.
120. Ingmar Pörn, *Action Theory and Social Science. Some Formal Models*. 1977.
121. Joseph Margolis, *Persons and Minds. The Prospects of Nonreductive Materialism*. Boston Studies in the Philosophy of Science, Volume LVII. 1977.
122. Jaakko Hintikka, Ilkka Niiniluoto, and Esa Saarinen (eds.), *Essays on Mathematical and Philosophical Logic*. 1978.
123. Theo A. F. Kuipers, *Studies in Inductive Probability and Rational Expectation*. 1978.

124. Esa Saarinen, Risto Hilpinen, Ilkka Niiniluoto, and Merrill Provence Hintikka (eds.), *Essays in Honour of Jaakko Hintikka on the Occasion of His Fiftieth Birthday*. 1978.
125 Gerard Radnitzky and Gunnar Andersson (eds.), *Progress and Rationality in Science*. Boston Studies in the Philosophy of Science, Volume LVIII. 1978.
126. Peter Mittelstaedt, *Quantum Logic*. 1978.
127. Kenneth A. Bowen, *Model Theory for Modal Logic. Kripke Models for Modal Predicate Calculi.* 1978.
128. Howard Alexander Bursen, *Dismantling the Memory Machine. A Philosophical Investigation of Machine Theories of Memory*. 1978.
129. Marx W. Wartofsky, *Models: Representation and the Scientific Understanding*. Boston Studies in the Philosophy of Science, Volume XLVIII. 1979.
130. Don Ihde, *Technics and Praxis. A Philosophy of Technology*. Boston Studies in the Philosophy of Science, Volume XXIV. 1978.
131. Jerzy J. Wiatr (ed.), *Polish Essays in the Methodology of the Social Sciences*. Boston Studies in the Philosophy of Science, Volume XXIX. 1979.
132. Wesley C. Salmon (ed.), *Hans Reichenbach: Logical Empiricist*. 1979.
133. Peter Bieri, Rolf-P. Horstmann, and Lorenz Krüger (eds.), *Transcendental Arguments in Science. Essays in Epistemology*. 1979.
134. Mihailo Marković and Gajo Petrović (eds.), *Praxis. Yugoslav Essays in the Philosophy and Methodology of the Social Sciences*. Boston Studies in the Philosophy of Science, Volume XXXVI. 1979.
135. Ryszard Wójcicki, *Topics in the Formal Methodology of Empirical Sciences*. 1979.
136. Gerard Radnitzky and Gunnar Andersson (eds.), *The Structure and Development of Science*. Boston Studies in the Philosophy of Science, Volume LIX. 1979.
137. Judson Chambers Webb, *Mechanism, Mentalism, and Metamathematics. An Essay on Finitism*. 1980. (Forthcoming.)
138. D. F. Gustafson and B. L. Tapscott (eds.), *Body, Mind, and Method. Essays in Honor of Virgil C. Aldrich*. 1979.
139. Leszek Nowak, *The Structure of Idealization. Towards a Systematic Interpretation of the Marxian Idea of Science*. 1979.
140. Chaim Perelman, *The New Rhetoric and the Humanities. Essays on Rhetoric and Its Applications*. 1979.
141. Wlodzimierz Rabinowicz, *Universalizability. A Study in Morals and Metaphysics*. 1979.

MIX
Papier aus verantwortungsvollen Quellen
Paper from responsible sources
FSC® C105338

If you have any concerns about our products,
you can contact us on
ProductSafety@springernature.com

In case Publisher is established outside the EU,
the EU authorized representative is:
Springer Nature Customer Service Center GmbH
Europaplatz 3, 69115 Heidelberg, Germany

Printed by Libri Plureos GmbH
in Hamburg, Germany